T0184100

SpringerBriefs in History of Science and Technology

Series Editors

Gerard Alberts, University of Amsterdam, Amsterdam, The Netherlands

Theodore Arabatzis, University of Athens, Athens, Greece

Bretislav Friedrich, Fritz Haber Institut der Max Planck Gesellschaft, Berlin, Germany

Ulf Hashagen, Deutsches Museum, Munich, Germany

Dieter Hoffmann, Max-Planck-Institute for the History of Science, Berlin, Germany

Simon Mitton, University of Cambridge, Cambridge, UK

David Pantalony, University of Ottawa, Ottawa, ON, Canada

Matteo Valleriani, Max-Planck-Institute for the History of Science, Berlin, Germany

More information about this series at http://www.springer.com/series/10085

Hubert Goenner · Giuseppe Castagnetti

Establishing Quantum Physics in Berlin

Einstein and the Kaiser Wilhelm Institute
for Physics, 1917–1922

 Springer

Hubert Goenner
Fakultät für Physik
Institut für Theoretische Physik
(emeritus) Georg-August-Universität
Göttingen
Göttingen, Germany

Giuseppe Castagnetti (Deceased)
Research Scholar
Berlin, Germany

ISSN 2211-4564   ISSN 2211-4572   (electronic)
SpringerBriefs in History of Science and Technology
ISBN 978-3-030-63121-5   ISBN 978-3-030-63122-2   (eBook)
https://doi.org/10.1007/978-3-030-63122-2

© The Author(s), under exclusive license to Springer Nature Switzerland AG 2020
This work is subject to copyright. All rights are solely and exclusively licensed by the Publisher, whether the whole or part of the material is concerned, specifically the rights of translation, reprinting, reuse of illustrations, recitation, broadcasting, reproduction on microfilms or in any other physical way, and transmission or information storage and retrieval, electronic adaptation, computer software, or by similar or dissimilar methodology now known or hereafter developed.
The use of general descriptive names, registered names, trademarks, service marks, etc. in this publication does not imply, even in the absence of a specific statement, that such names are exempt from the relevant protective laws and regulations and therefore free for general use.
The publisher, the authors and the editors are safe to assume that the advice and information in this book are believed to be true and accurate at the date of publication. Neither the publisher nor the authors or the editors give a warranty, expressed or implied, with respect to the material contained herein or for any errors or omissions that may have been made. The publisher remains neutral with regard to jurisdictional claims in published maps and institutional affiliations.

This Springer imprint is published by the registered company Springer Nature Switzerland AG
The registered company address is: Gewerbestrasse 11, 6330 Cham, Switzerland

Contents

Chapter 1
Einstein Comes to Berlin

Abstract This chapter shows how a group of prominent physicists, with whom Einstein had interacted scientifically during previous years, promoted his call to Berlin, and the establishment of a research institute under his directorship in the expectation that Einstein and the institute would take the lead in advancing quantum physics in its early phase. It explores the expectations placed upon him in the context of his achievements, the establishment of the Kaiser Wilhelm Society, and the state of physics during the first decades of the twentieth century.

Keywords Kaiser-Wilhelm-Institut für physikalische Forschung · Hugo Andres Krüss · Fritz Haber · Walter Nernst · Max Planck · Ruebens · Emil Warburg

1.1 Relations Between Einstein and Berlin Physicists

The call of Einstein to Berlin was the result of efforts made by several members of the Berlin scientific community over a period of years and must be seen in the context of institutional as well as purely scientific interests.[1]

There is evidence that as early as 1910 Walther Nernst had thought of a way to bring the thirty-one-year-old extraordinary professor of theoretical physics from the University of Zurich to the center of physics research in Germany (Fölsing 1993, 337; Kormos Barkan 1999, 183, note 3). However, the first concrete step in this direction was made, as far as we know, by Emil Warburg in April 1912 when he offered Einstein, visiting Berlin at the time, a position at his institute, the Physikalisch-Technische Reichsanstalt (Imperial Institute for Physics and Technical Standards).[2] In fact, the scientific relations between Einstein and some senior Berlin physicists dated some years back. As early as 1906 a lively correspondence began with Max Planck, which was later followed by others. Einstein had also met several German

---

[1]On Einstein's call to Berlin, see Schulmann (1995). Our succinct biographical notes concerning Einstein are based on the detailed narratives given by Fölsing (1993), Frank (1949), Hermann (1994), Seelig (1960). As to the course of Einstein's scientific activities, we refer to Pais (1982).

[2]Einstein (1993b, 415, 480, 511). On the Physikalisch-Technische Reichsanstalt, see Cahan (1989).

© The Author(s), under exclusive license to Springer Nature Switzerland AG 2020
H. Goenner and G. Castagnetti, *Establishing Quantum Physics in Berlin*,
SpringerBriefs in History of Science and Technology,
https://doi.org/10.1007/978-3-030-63122-2_1

colleagues at congresses or private meetings between 1909 and 1911, so that when he traveled to Berlin in April 1912 in order to "talk shop with various people"[3] he was already an acknowledged figure. The program of his visit included meetings with Fritz Haber, Nernst, Planck, Heinrich Rubens, Warburg, and the young astronomer Erwin Freundlich (Einstein 1993a, 581; Einstein 1993b, 467). It is worth expounding in some detail the scientific aspects of these personal relations, bearing in mind that their intensity and cordiality were at that time anything but obvious.

Haber[4] was director of the newly opened Kaiser-Wilhelm-Institut für physikalische Chemie und Elektrochemie (Kaiser Wilhelm Institute for Physical Chemistry and Electrochemistry) and professor of physical chemistry at the University of Berlin. He was known worldwide for his and Bosch's method for the synthesis of ammonia. In addressing problems in his discipline, Haber paid keen attention to the development of physical theories concerning the constitution of matter, like the new quantum theory (Bonhoeffer 1953, 5). He first met Einstein in September 1911 at the congress of the Gesellschaft Deutscher Naturforscher und Aerzte (Society of German Scientists and Physicians) in Karlsruhe. He felt so stimulated by the discussions he had there with Einstein that he felt compelled to write to him afterwards that "no one has taught me more than you," and to ask him to "continue teaching me through your criticism."[5] The ensuing correspondence dealt with solid-state theory, in particular Haber's hypothesis concerning the relative proper frequencies of electrons and atoms in crystals. Haber had also worked on the energy balance in chemical reactions and suggested a possible connection with the quantum hypothesis, namely with the quantized energy carried by electrons in such reactions. Einstein, however, did not believe Haber's results, and in general had mixed feelings towards Haber's work in physics, although he granted him an "abundance of ideas."[6] Haber had also shown interest in the precision electrometer that Einstein had developed with Richard Conrad and Paul Habicht (Einstein 1993b, 383).

Nernst[7] was professor of physical chemistry at the University of Berlin and director of the corresponding institute. He enjoyed international reputation for his contributions to the new developments in physical chemistry and for his heat theorem. For several years, Nernst had been the only major scientist to address the theoretical and experimental consequences for physical chemistry of the concept of the quantization

---

[3] Einstein to Besso, 26 March [1912], in Einstein (1993b, 437, doc. 377). For quotations from texts published in the *Collected Papers of Albert Einstein* we adopt the translations, with slight modifications, from Einstein (1995b, 1998a, 2004a). In all other cases, the translations are by the authors.

[4] On Haber, see Bonhoeffer (1953), Stoltzenberg (1994), Szöllösi-Janze (1998). Unless otherwise noted, bio-bibliographical information is drawn from the *Poggendorffs* bibliographies (Weinmeister 1925–1926; Stobbe 1936–1940; Zaunick and Salié 1956–1962).

[5] Haber to Einstein, 19 December 1911, in Einstein (1993b, 337, doc. 329).

[6] Einstein to Heinrich Zangger, undated [20 December 1911], in Einstein (1993b, 379, doc. 330); see also Einstein (1993b, docs. 308, 364, 368).

[7] On Nernst, see Kormos Barkan (1999), Bartel and Huebener (2007), Bodenstein (1942), Hiebert (1978).

of energy introduced by Einstein's paper, which explained the specific heat of solids by the use of Planck's black-body radiation formula.[8] And it was mainly because of Nernst's publications that Einstein's paper became known among physicists. Einstein's theoretical result, namely that specific heats vanish at absolute zero, confirmed Nernst's heat theorem. Hence Nernst embarked on an extensive program of measurements of specific heat of solids at low temperature, a program in which Rubens and some of his collaborators also took part. In 1911 this research led Nernst to confirm Einstein's predictions concerning the specific heat of solids at low temperatures ·except for certain discrepancies. Meanwhile, in March 1910 Nernst visited Einstein in Zurich and was so impressed that he called him a "Boltzmann redivivus."[9] It was on Nernst's initiative that the First Solvay Congress was convened in autumn 1911 in Brussels, at which a selected group of the most prominent European physicists of the time discussed developments in radiation theory in light of the quantum hypothesis.[10] Of course, the young and still little-known Einstein was also invited and, on that occasion, made his official entrée into the inner circle of the international physics community. Whereas Nernst was almost enthusiastic in his appreciation of Einstein during this period, for his part Einstein was more skeptical about Nernst's qualities as a theoretician[11] and criticized his repeated attempts to derive the heat theorem from thermodynamics. He later insisted that quantum theory would be necessary in order to do this.

Planck[12] was professor of theoretical physics at the University of Berlin and a widely recognized authority among German physicists. In June 1912, shortly after Einstein's visit, he was appointed as one of the four "Permanent Secretaries" of the Preussische Akademie der Wissenschaften (Prussian Academy of Sciences), that is, he became a leading figure of the most prestigious scientific institution in the country. Although Planck became co-editor of the *Annalen der Physik* only in 1906, in view of his function as scientific adviser to the journal since 1895, he had probably already played a role in the publication of Einstein's first papers in 1901. In autumn 1905, immediately after the publication of Einstein's "Zur Elektrodynamik bewegter Körper," expounding the theory of special relativity, (Einstein 1905) Planck had given a report on it to the "physical colloquium" at the University (Fölsing 1993, 227). Planck became an "enthusiastic relativist" and was "instrumental in securing the swift acceptance" of special relativity among physicists (Heilbron 1986, 28). In addition to his own reflections on the subject, between 1905 and 1914 Planck also supervised seven doctoral dissertations directly or indirectly concerning special relativity (Hoffmann 1984, 60). In spring 1906 at the latest, he began to correspond with

---

[8]Einstein (1907). For Nernst's contribution to quantum physics and his relationship with Einstein, see Kormos Barkan (1999, 164–207). See also Einstein (1993b, docs. 199, 270, 364, 366, 384).

[9]Nernst to Arthur Schuster, 17 March 1910, quoted in Kormos Barkan (1999, 183).

[10]Kormos Barkan (1999, 181–207), Mehra and Rechenberg (1982, 127–136). For the proceedings, see Langevin and de Broglie (1912).

[11]Privately, Einstein considered Nernst "a fabulous technician." Einstein to Zangger, 20 May 1912, in Einstein (1993b, 467, doc. 398).

[12]On Planck, see Born (1948–1949), Heilbron (1986).

Einstein on various problems at the forefront of physics research (Einstein 1993b, docs. 36, 47, 64, 118, 160, 161, 172, 211). In addition to the question of an empirical verification of special relativity, black-body radiation and Planck's first quantum hypothesis were also discussed from the very beginning of their exchange. Einstein met Planck as well as Rubens, Arnold Sommerfeld, and Wilhelm Wien, for the first time in September 1909, at the congress of the Gesellschaft Deutscher Naturforscher und Aerzte in Salzburg where he gave a lecture on the nature of radiation (Einstein 1909, 1993b, 227). Planck and Einstein differed on Einstein's hypothesis of the corpuscular nature of electromagnetic waves.[13] Their discussion continued unabated during the First Solvay Congress in 1911, at which Planck presented his second quantum hypothesis and strove to shift the emphasis from quantization of energy to the quantization of action in phase space (Planck 1914). Privately, Einstein rejected Planck's second theory as "almost worthless."[14] While "Planck had the very greatest admiration for Einstein's work, which he liked to compare with Copernicus," (Heilbron 1986, 31) Einstein tended to consider Planck a stubborn conservative because of his skepticism towards new physical theories.[15]

Rubens was professor of experimental physics at Berlin University and director of the university's Physical Institute.[16] For many years he had worked systematically on Maxwell's theory of electromagnetism and contributed to the emergence of quantum physics. Under Rubens's supervision, a doctoral thesis confirming special relativity had been completed in 1909 (Einstein 1993b, 186–187, note 9). As mentioned, Einstein and Rubens met shortly thereafter at the congress in Salzburg. At the time, Rubens was working with Nernst on the experimental proof of Einstein's predictions concerning the specific heat of solids.[17] In January 1912, Einstein and Rubens corresponded about Rubens's measurements of the wavelengths of residual rays, from which it was thought that the optical characteristic frequencies of crystals could be obtained. This in turn was needed to verify Einstein's formula for the temperature dependency of the specific heat of a solid. Einstein at first misinterpreted Rubens's results and wrote a paper on the subject but withdrew it due to Rubens's criticism.[18]

Warburg[19] was the President of the Physikalisch-Technische Reichsanstalt, one of the major institutes for experimental physics in Germany. Since at least 1907, following the formulation of the light quantum hypothesis by Einstein, Warburg had been doing theoretical work concerning the energy balance of photochemical reactions, besides continuing investigations into chemical reactions caused by electric discharges in gases. Einstein met Warburg for the first time in autumn 1911 at the

---

[13]Einstein to Johannes Stark, 31 July 1909, in Einstein (1993b, 202, doc. 172).

[14]Einstein to Wien, 17 May 1912, in Einstein (1993b, 464, doc. 395).

[15]See, e.g., Einstein (1993b, 349, 588–589).

[16]On Rubens, see Kangro (1975).

[17]Einstein to Jakob Laub, 16 March 1910, in Einstein (1993b, 232, doc. 199).

[18]For a reconstruction of this controversy, see Kox (1995); see also Einstein (1993b, docs. 313, 331, 343, 344, 354, 377).

[19]On Warburg, see Ramser (1976).

Solvay Congress, during which they occasionally discussed photochemistry. Einstein disagreed with some of Warburg's conclusions concerning the photochemical energy balance.[20] It was allegedly during these conversations that Einstein found a theoretical proof, based on thermodynamical reasoning, of the quantum nature of light absorption in photochemical phenomena, which became known as the law of photochemical equivalence. A few weeks later, he published a paper on the subject (Einstein 1912). Warburg, on the other hand, started a series of experiments lasting for several years that indeed verified Einstein's law of photochemical equivalence under certain conditions. He occasionally reported his results to Einstein and the two also discussed the subject during their meeting in April 1912 in Berlin. It is quite clear that their common interest in the photochemical energy balance, and the scientific exchange that stemmed from it, were behind Warburg's decision to invite Einstein to join his institute. In all likelihood, the invitation was also part of Warburg's plan to reform the Reichsanstalt and to broaden the scope of its research, in order to bridge the increasing gap between the traditional concerns of the Reichsanstalt in technical physics and new developments in pure physics (Cahan 1989, 176–187).

Erwin Freundlich[21] became an assistant at the Sternwarte, the Berlin Astronomical Observatory, in 1910. In August 1911, he came into contact with Einstein, at the time professor at the German University in Prague, who was enquiring among astronomers about the possibility of empirically testing the influence of gravitation on the propagation of light as predicted by his relativistic theory.[22] In particular, one question raised was whether it would be possible to observe a deflection of light rays in the gravitational field of the sun.[23] Freundlich immediately took up "with passion" the verification of this phenomenon.[24] A long correspondence began, from which regrettably very few of Freundlich's letters still exist.[25] The two most probably met for the first time in April 1912 during Einstein's visit to Berlin. Freundlich was the first and for a long time "the only colleague in that profession [astronomy] until now to support [Einstein's] efforts effectively in the area of general relativity."[26] Not only did he work for years on several ways to test Einstein's theory, but it was also thanks to his publications that "the astronomers [...] started to show interest

---

[20] On this subject and on Einstein's relations with Warburg, see the editorial comment "Einstein on the Law of Photochemical Equivalence" in Einstein (1995a, 109–113). See also Einstein (1993b, docs. 308, 354, 362, 366, 385, 386).

[21] For bio-bibliographical information on Freundlich, see Forbes (1972). Exhaustive treatments of the relations between Einstein and Freundlich are given in Hentschel (1992, 1994), see also Pyenson (1985, 228–236).

[22] Käthe Freundlich to Lewis Pyenson, 29 April 1973, copy in Albert Einstein Archives [hereafter AEA], 11–241; Pyenson (1985, 230).

[23] For the history of the research concerning the gravitational redshift, see Hentschel (1998), Earman and Glymour (1980).

[24] Einstein to Zangger, 20 September 1911, in Einstein (1993b, 325, doc. 286).

[25] For the period 1911–1921, see the letters published in Einstein (1993b, 1998b, 2004b, 2006, 2009).

[26] Einstein to Sommerfeld, 2 February 1916, in Einstein (1998b, 256, doc. 186).

in the important question about the bending of light rays."[27] Freundlich "became Einstein's mouthpiece among astronomers," (Hentschel 1994, 154) which was not without consequences for his academic career, since he often confronted hostility or at least strong skepticism. In December 1913, shortly after his election to the Prussian Academy of Science but before his move to Berlin, Einstein asked Planck to support Freundlich's request for financial help from the Academy for an expedition to Crimea, to observe the stars near the sun during the solar eclipse of August 1914. The expedition was so important to Einstein that he even envisaged borrowing money from a private donor or contributing from his own savings.[28]

Einstein also had important links to Berlin outside the scientific community. Some of his relatives lived in this city, among them his cousin Elsa Einstein. During his visit in April 1912, Einstein fell in love with her; they married in 1919. For the time being though, this love affair was not enough to convince Einstein to move to Berlin, as he hinted in a letter to Elsa written immediately after his return home, in which he complained: "It is such a pity that we don't live in the same town. The chances of my getting a call to Berlin are, unfortunately, rather slight, as I must admit to myself when I think about it clearly."[29]

It is not clear why Einstein declined Warburg's offer. Although in the quoted letter to his cousin he regrets that he had a slim chance of being called to Berlin, we know that he himself did not want to take Warburg's offer into consideration nor the call to Vienna received at about the same time, allegedly because he did not want to behave incorrectly towards the Eidgenossische Technische Hochschule (ETH, Federal Polytechnic) of Zurich. Einstein had just been appointed professor of theoretical physics at the ETH in January 1912 and had moved from Prague to Zurich the following summer (Einstein 1993b, 630–631). In addition, a later allusion by Haber[30] to Einstein's refusal gives some reason to believe that he probably did not want to be tightly bound to an institute like the Physikalisch-Technische Reichsanstalt, whose institutional obligations included standard measurements for technical applications in a great range of fields. After all, Einstein was satisfied with the institute "in [his own] head" and needed "in addition at most a few books."[31]

---

[27]Einstein to Freundlich, undated [August 1913], in Einstein (1993b, 550, doc. 468), Hentschel (1992, 26).

[28]Einstein to Freundlich, 7 December 1913, in Einstein (1993b, 581, doc. 492). In fact, the Academy granted 2,000 marks or about one third of the expedition costs on the recommendation of Planck, Rubens and Karl Schwarzschild. The director of the Sternwarte, Hermann Struve, also supported the project, although he explicitly doubted that the observations would provide conclusive evidence (see Freundlich's application to the Academy, 7 December 1913, Archiv der Berlin-Brandenburgischen Akademie der Wissenschaften [hereafter AAdW], Berlin, II–VII, Bd. 157, p. 137; minutes of the meeting of the mathematical-physical class, 11 December 1913, AAdW, II–V, Bd. 132, p. 117). See also Hentschel (1992, 51–52).

[29]Albert Einstein to Elsa Einstein, undated [30 April 1912], in Einstein (1993b, 456, doc. 389).

[30]Haber to Hugo Andres Krüss, 4 January 1913, in Einstein (1993b, 511, doc. 428).

[31]"Statement of Reasons for Leaving Prague," 3 August 1912, in Einstein (1993b, 499, doc. 414).

1.2 Einstein's Recruitment

Warburg's offer to Einstein was probably made on his own initiative and not in agreement with the other leading physicists. Indeed, the name Einstein had not even been mentioned during a meeting concerning the foundation of an institute for physical research held just a couple of months earlier, in February 1912.[32] In any case, after Warburg's first attempt other members of Berlin's scientific establishment joined the effort to recruit Einstein.

Eight months after Einstein's visit, in a letter[33] to an official in the Prussian Ministry of Education, Hugo Andres Krüss, Haber took up a suggestion that Krüss had made some time earlier, namely "whether there could not be created a position for this extraordinary man in the institute in my charge." Haber not only strongly backed this idea but had meanwhile discussed it with Krüss's superior in the Ministry, *Ministerialdirektor* Friedrich Schmidt-Ott, and also with the financier Leopold Koppel who was willing to support Einstein's recruitment to Berlin financially.[34] Following a remark by Schmidt-Ott that the theorist Einstein would not need an institute with laboratories, Haber agreed that Einstein be given a special status in his institute: "Even a theoretical physicist of his orientation is in need of certain resources in order to study experimentally one topic or another from time to time, or to have it studied by an assistant or collaborator" (Einstein 1993b, 511). In his preliminary calculations, Haber estimated that a single expenditure of 50,000 marks for scientific equipment would be sufficient and a yearly salary of 15,000 marks adequate for Einstein. This was a high salary for a young professor. Interestingly enough, in support of the plan Haber argued that Einstein was to be treated like a scientific successor to Jacobus van't Hoff, the famous physical chemist who had held the first research professorship established at the Prussian Academy of Science. Before any further action was taken, it was considered appropriate that the colleague scientifically closest to Einstein, the theoretical physicist Planck, should be approached (Einstein 1993b, 513–514; vom Brocke 1990, 87). Apparently, Planck was not yet involved and Haber's initiative, like Warburg's, was still a personal one, not a concerted action.[35]

Through the interaction between Haber, Planck, and Nernst during the first half of 1913, the original Krüss-Haber plan was changed dramatically (Einstein 1993b, 529, note 1). It was no longer envisioned that Einstein would become a researcher

---

[32] Minutes of the meeting of 5 February 1912, GStA, I. HA, Rep. 76 V c, Sekt. 2, Tit. 23, Litt. A, Nr. 116, pp. 16–17.

[33] Haber to Krüss, 4 January 1913, in Einstein (1993b, 510–514, notes and doc. 428).

[34] In 1905, Koppel contributed considerable funds to a Koppel-Stiftung (Koppel Foundation) aimed at supporting philanthropic and educational schemes under the patronage of Kaiser Wilhelm II. The Koppel-Stiftung was the largest contributor to the Kaiser-Wilhelm-Gesellschaft (see below, Sect. 1.4) from its foundation and single-handedly funded Haber's KWI für physikalische Chemie und Elektrochemie. On Koppel and the Koppel-Stiftung, see vom Brocke (1990, 98–102).

[35] In the correspondence between Albert Einstein and his cousin Elsa Einstein there is evidence that she also personally recommended Einstein to Haber (Einstein 1993b, 544–546, docs. 465, 466). It is not known what kind of relationship existed between Elsa Einstein and Haber at that time, but Elsa's intervention was not a decisive factor in Haber's initiative.

with special status in Haber's institute but rather a member of the scientific Olympus, (Einstein 1993b, 259) the Prussian Academy of Science. On 12 June 1913 Planck read to the physical-mathematical class of the Academy a proposal he had formulated himself, which was also signed by Nernst, Rubens, and Warburg, to elect Einstein as a member.[36] On that occasion Nernst informed the Academy members that half of Einstein's exceptional yearly salary of 12,000 marks would be provided by Koppel (Einstein 1993b, 529, note 2). The usual annual honorary remuneration for members was 900 marks. From 12 to 14 July 1913, Planck and Nernst visited Einstein in Zurich and offered him a membership in the Academy with a joint honorary professorship at the University of Berlin under the conditions mentioned. He would be free to continue his research with the right but no obligation to teach, that is, he would have the same status as van't Hoff, as Haber had suggested. Although the establishment of a KWI for physical research under Einstein's directorship was not mentioned in the documents of the Academy, references to it in Einstein's letters indicate that the project had certainly been discussed, at least in general terms, during the meeting in Zurich (Einstein 1993b, docs. 451, 453, 478, 482, 509).

This time the conditions were convincing and Einstein accepted the offer. In his private correspondence he mentioned the wish to be closer to his cousin Elsa as one of the main reasons for his decision.[37] Einstein's relationship with his wife Mileva had become so wretched (Einstein 1993b, docs. 489, 497, 498) that he intentionally headed for a situation that would make things worse, apparently without being aware of the consequences, as later became evident. Einstein's private remarks are not always to be taken at face value, though. Beside the financial aspects of the new position, the decisive reasons for Einstein's move to Berlin were probably—as he also hinted—the dispensation from teaching obligations, the possibility of working with junior collaborators in his own institute, and the prospect of fruitful scientific exchange with his senior colleagues (Einstein 1993b, docs. 453, 454, 455, 482, 484, 488, 513).

The physical-mathematical class of the Prussian Academy had already elected Einstein on 3 July 1913, that is, before Nernst's and Planck's visit to Zurich, with 21 votes to one, six members being absent.[38] The election was confirmed at the plenary session of the Academy on 24 July 1913 with 44 votes to two. This time, two members abstained and eleven were absent.[39] With his formal acceptance of the nomination Einstein announced his intention to move to Berlin at the beginning of April 1914. He actually arrived on 29 March 1914 (Einstein 1993b, 582; Einstein 1998b, 11). Since the new institute had not yet been established, Einstein was given an office in Haber's institute (Einstein 1993b, 604; Einstein 1998b, 13).

---

[36]Kirsten and Treder (1979, 95–97, vol. 1), Einstein (1993b, 526–528, doc. 445). Haber was not among the signers because he became a member of the Academy in 1914.

[37]Einstein (1993b, docs. 454, 465, 488, 509), Einstein (1998b, docs. 2, 94). This question is also discussed in Fölsing (1993, 374–378).

[38]Minutes of the meeting, AAdW, II–V, Bd. 132, pp. 93–94; (Kirsten and Treder 1979, 98, vol. 1).

[39]Minutes of the meeting, AAdW, II–V, Bd. 89, pp. 57–59.

1.3 What Was Expected from Einstein?

For what reasons did the most prominent physicists and chemists in Berlin so dearly want Einstein to become their colleague? In order to answer this question we must take into consideration the state of physics during the first decades of the twentieth century.[40] The crisis in theoretical physics had not yet subsided. It is true that special relativity and classical mechanics had already been reconciled with one another in a new mechanics (Laue 1911). However, quantum physics was a mixture of theoretical hypotheses and a multitude of little-understood empirical findings related to Planck's energy quantum. Both the wave theory of light as described by Maxwell's equations and Bohr's atomic model still presented formidable obstacles. With the concept of ether banished, what supports the oscillations of light waves? What about Einstein's photons? The atomic model that Bohr suggested in 1913 was in conflict with Maxwell's electrodynamical theory: in their accelerated motion around the nucleus electrons should have radiated but did not. Everyone in the scientific community was aware of this critical situation. Einstein still described it in dramatic terms in his "Autobiographical Notes": "It was as if the ground had been pulled out from under one, with no firm foundation to be seen anywhere upon which one could have built" (Einstein 1979, 42–43). Haber, in a memorandum for the foundation of the Kaiser-Wilhelm-Institut für physikalische Forschung, used the following words:

> With each month more evidence accumulates that the fundamental concept introduced by Planck into theoretical physics with the so-called elementary quantum of action is indispensable for [understanding] all the processes of the molecular world. But, until now, it was not possible to reconcile this concept in a satisfactory manner with the concepts of classical theoretical physics. Physics therefore needs very urgently new experimental research in order to obtain a usable basis for the physical understanding of nature. The importance of this task cannot be overestimated.[41]

As to Einstein's role, we noted already that he had interacted scientifically with all the main Berlin physicists during the previous years. The range of fields around which they had met ran from relativity to quantum theory, from thermodynamics to photoelectricity and photochemistry. Nernst "surely hoped that Einstein would continue to be as close to experiment as he had been" (Kormos Barkan 1999, 186) up to that time, thus helping him in the effort to integrate quantum theoretical concepts in a molecular theory of chemical processes. Haber's hope was that Einstein could play a similar role for theoretical chemistry as van't Hoff, who had introduced to it thermodynamics. In Haber's opinion, Einstein was precisely the person to achieve a similar success by putting to use "the radiation theory and electromechanics" to advance chemistry. By "electromechanics" we may understand "the theory of the electron" or, more broadly, the interaction of atoms and electromagnetic radiation,

---

[40]For the history of physics of this period, see Mehra and Rechenberg (1982), Jungnickel and McCormmach (1986, vol. 2), Kragh (1999).

[41]"Gegen Ende des vorigen Jahrhunderts ..." Staatsbibliothek Berlin [hereafter SBB], Acta Preussische Staatsbibliothek, Generaldirektion, Kaiser-Wilhelm-Institute XXVI. The typewritten document is not signed but, on 25 April 1914, Krüss added: "received from privy councilor Haber."

the quantum mechanical nature of which was largely unexplored and not well understood. "This fundamental task—Haber concluded—can be incomparably advanced by having Mr. Einstein join our institute."[42]

The expectations indirectly expressed by Planck in the proposal of June 1913 for Einstein's membership in the Academy were much higher, and based upon a better knowledge of Einstein's scientific production (Kirsten and Treder 1979, 95–97, vol. 1; Einstein 1993b, 526–528). Planck, of course, considered the theory of special relativity as a fundamental achievement, but he also thought that its consequences were "at the very limit of the measurable" and consequently of lesser interest than Einstein's contributions to other fields of physics. In his words "much more significant [than the theory of special relativity] for experimental physics is his tackling of other problems that are at the moment at the center of attention." Among the contributions Planck enumerated the first was the application of the "quantum hypothesis" to the "atomic and molecular motions" and its consequences for the "development of the new kinetic *atomistics*."[43] Second, he mentioned the "verifiable relationships" that Einstein had established between the "quantum hypothesis" and the "lightelectric and photoelectric effects," and, finally, the "kinship" Einstein had pointed out "between the constants of elasticity and those of the optical proper vibrations of crystals." Not only in the "formulation and critique of new hypotheses" should Einstein be considered a "master," according to Planck, but also in the "deepening of classical theory." Here, Einstein's "preferred field of research is the kinetic theory of matter and its relation to the fundamental laws of thermodynamics," a field in which his contributions had opened up new directions for experimental research. While remarking that "among the big problems which are so abundant in modern physics, there is hardly one in which Einstein did not take a position in a remarkable manner," Planck could kindly excuse as juvenile "speculations" Einstein's "light quantum hypothesis" or skeptically report on his attempts to develop "a new theory of gravitation." For Planck, the most important aspect of Einstein's scientific personality was his "talent for quickly getting to the bottom of other scientists' newly emerging views and assertions, and for assessing their relationship to each other and to experience with surprising certainty."

Thus, it seems clear to us that Einstein was brought to Berlin in order to bring fresh inspiration for the further theoretical and experimental development of quantum physics and its applications to material science and physical chemistry, and not at all to work on a generalization of special relativity towards a relativistic theory of gravitation. Einstein must have understood this; in a letter to his friend Michele Besso he comments: "The fraternity of physicists behaves rather passively with respect to my paper on gravitation. [...] Laue is not open to the fundamental considerations, and neither is Planck."[44] Of course, in speaking of quantum theory, we must not understand the term as we do today. For Planck, the "core of the quantum hypothesis as well as of Nernst's heat theorem" consisted merely in the thermodynamical

---

[42]Haber to Krüss, 4 January 1913, in Einstein (1993b, 511, doc. 428).

[43]Emphasis in the original.

[44]Einstein to Besso, undated [January 1914], in Einstein (1993b, 588–589, doc. 499).

principle that the entropy of a system does not diminish without boundary but becomes zero at the minimum temperature (Planck 1913, vii). Nernst seems to have wavered between considering the quantum, at some moments, a mere "calculation rule"[45] and, at others, a very powerful instrument for understanding physical-chemical phenomena that should be incorporated in a new molecular theory of the solid state (Nernst 1913, 254–268; Kormos Barkan 1999, 166, 174–179).

Even if the establishment of a physics institute had been considered during the Zurich meeting in July 1913, the Berliners did not link the expectations they placed in Einstein's "talent" to the project of giving him a leading function in such an institute, at least not at that time. Until early 1914 they were undecided as to the organization, the purpose, and even the directorship of the new institute they envisaged. In autumn 1913 Einstein was informed that the matter had been postponed until his arrival and he did not hear any more about it until then (Einstein 1993b, docs. 482, 509, 513). Only on 9 January 1914 did a meeting concerning the planned institute take place between Planck, Nernst, Haber, the representatives of the Prussian Ministry of Education Schmidt-Ott and Krüss, and the financier Koppel. On that occasion, Planck still pleaded to wait until Einstein had adjusted to the circumstances in Berlin and suggested instead supporting him with "astronomical works" in the field of research in which he was primarily engaged at the time, namely gravitation theory. Nernst had a totally different opinion and maintained that it was "time to attack the theory of the solid state with massive support," pointing to Max von Laue's and William Henry Bragg's research in unraveling the structure of crystalline bodies using X-ray scattering. In order to carry out this "attack" he suggested two alternative "ways": either the establishment of a small institute under the direction of Hans Geiger, a physicist working at the Physikalisch-Technische Reichsanstalt, that would undertake research in that field as well as on "radiology," or the appointment of a committee including Planck, Einstein, Haber, Warburg, Beckmann, Rubens, Laue, and himself, which would allocate funds for research done by others: "the committee has the [financial] means. Cheapest way. It is likely that the topic will be exhausted within four to five years."[46]

It was at this stage that a way was found to integrate Einstein into the structure of a new institute that would tackle the problems singled out by Nernst and Haber. At the beginning of February 1914, Haber, Nernst, Planck, Rubens, and Warburg submitted to the Prussian Government, the Kaiser-Wilhelm-Gesellschaft zur Förderung der Wissenschaften (Kaiser Wilhelm Society for the Advancement of Science) and the Koppel-Stiftung a proposal for the foundation of a "Kaiser-Wilhelm-Institut für physikalische Forschung" (for physical research, hereafter KWI für Physik) with Einstein as the "permanent honorary secretary" of the scientific steering board.[47] As one of the "first tasks" of the new institute the initiators, following Nernst's sug-

---

[45]Nernst's quotation in Kormos Barkan (1999, 174).

[46]Minutes of the meeting, 9 January 1914, Geheimes Staatsarchiv Berlin [hereafter GStA], I. HA, Rep. 76 V c, Sekt. 2, Tit. 23, Litt. A, Nr. 116, pp. 18 r–v; Wendel (1975, 197).

[47]Kirsten and Treder (1979, 146–148, vol. 1), Wendel (1975, 198–199). The only original specimen of the proposal that we were able to find was sent by Nernst to Schmidt-Ott on 4 February 1914

gestion, set the "investigation of radioactive processes and further research into the structure of crystalline bodies with the help of X-rays." The study of radioactivity was considered one of the avenues that might possibly lead to an understanding of atomic structure.

1.4   A New Way of Organizing Research in Physics?

A further argument for our supposition that Einstein was recruited for his assumed ability to lead a coordinated research effort in what we now call the microscopic theory of matter follows from the particular manner in which the new institute was to operate. As the founding proposal of February 1914 remarked, the envisaged manner of organizing research was "almost new in physics" (Kirsten and Treder 1979, 147, vol. 1; Kant 1996, 228).

In the last decades of the nineteenth century, German universities and their institutes had been systematically expanded to "large-size enterprises" in order to satisfy the needs of a rapidly changing society.[48] Because of the social and technical development that had taken place in Germany in the second half of the century, scientific knowledge had become an integral part of the production processes. Not only the demand for technical staff in industry increased, but also the need for new products and for improving existing products and production techniques. As a consequence, the support for scientific research became a matter of public interest. The state was now willing to allocate more money for higher education. Larger buildings with up-to-date equipment were built especially for the institutes for chemistry and physics, and the number of staff increased.[49]

Nevertheless, the work organization in the new university institutes did not change dramatically. Their main tasks were still the transmission of knowledge to the students through lectures and practical training, as well as the assessment through examinations of the knowledge acquired, the correction of papers, the supervision of theses, and the like. Furthermore, a considerable part of a professor's income came from the fees that the students had to pay for almost every course, practicum, and examination. The ordinary professor, who as a rule was also the director of the institute, usually reserved for himself the mandatory courses, which were attended by a large numbers of students and therefore brought more money. These courses, though, were also more general in content and required more preparation time. The more specialized courses were instead given by subordinate staff, like extraordinary professors or *Privatdozenten*, who had the right to teach but no permanent position and also a lower

---

(GStA, I. HA, Rep. 76 V c, Sekt. 2, Tit. 23, Litt. A, Nr. 116, pp. 3–5). On the foundation of the KWI für Physik and its preliminaries, see Sect. 2.1 below.

[48] Vom Brocke (1990, 85–86), vom Brocke (1996, 4–6). For the history of the university institutes in Germany, see also Riese (1977), Scheuch and von Alemann (1978). In particular, for the physical research institutes, see Forman (1968, 59–125), Hermann (1978), Kant and Hoffmann (1981). On the history of large-scale scientific research in Germany, see Ritter (1992).

[49] See, e.g., Riecke (1905), Wiener (1906).

income. Concerning the research work, the institute's chief was a kind of monarch in whom all power was invested. He decided on the allocation of money, the research subjects, the use of the instruments, the work of the assistants, and so on. The *Privat-dozenten* needed the director's consent in order to pursue their research and thereby launch their own academic careers, which almost always required a call to another university. In practice the institute's director, because of his institutional obligations, had almost no time for research, while the junior staff had no liberty to do as they liked. Furthermore, due to the differences in professional perspectives and research programs, and also because of the increasing differentiation and specialization within the disciplines, teamwork was unusual. On the contrary, rivalries and conflicts were not infrequent. As a consequence university institutes were unable to keep up with the development of increasingly complex research requiring cooperation between scholars.[50]

The foundation of the Kaiser-Wilhelm-Gesellschaft (hereafter KWG) and its insti-tutes in 1911 was intended to remedy this situation.[51] The Kaiser Wilhelm Institutes (hereafter KWI) had as their only stated task the performance of research. The direc-tors and collaborating scientists were given better working conditions than those at the university, and also had no teaching obligations.[52] The leading scientists, the researchers at the intermediate level, and the technical staff had permanent or long-term positions allowing them to work with continuity. The laboratories were generously equipped with more and more modern instruments. Furthermore, before the economic crisis of the early twenties, the institutes of the KWG received financial support not only from the state, which initially provided only the director's salary, but also from private donors, especially from industrial and financial companies, both directly and through the KWG, and thus had a larger budget than university institutes.[53]

Concerning the organization of personnel, the "monarchic" character was accen-tuated in comparison to that of a university institute. Following what became known as the "Harnack-Prinzip," a Kaiser Wilhelm Institute was built around a strong sci-entific personality chosen because of his outstanding results and promising research program. In order to realize this program, the statute of a KWI guaranteed the direc-tor the disposal over staff and the necessary financial means (vom Brocke 1996,

---

[50]Forman (1968, 66–79, 92–100), Hermann (1978), Burchardt (1988), vom Brocke (1996, 8–9). For contemporary references on tasks and organization of university institutes, see, e.g., Festschrift (1909, 24–69), Ramsauer (1913), von Staa (1930). On income and tuition, see also Jastrow (1930).

[51]For the history of the Kaiser-Wilhelm-Gesellschaft, see Burchardt (1975), Vierhaus and vom Brocke (1990), vom Brocke and Laitko (1996), Wendel (1975), Wendel (1984). The Kaiser-Wilhelm-Gesellschaft was the predecessor of the present Max-Planck-Gesellschaft.

[52]In fact, all the directors of the KW Institutes were also university professors, though not always full professors. It goes beyond the scope of this study to examine how teaching and research work were actually integrated. The concordant evidence is that teaching and institutional obligations were seen as hampering factors.

[53]On the foundation, structure, and financing of the KWG, see vom Brocke (1990).

18–20).[54] On the one hand, there was no more reason for rivalry at the higher levels; on the other though, the director had to be a person capable of organizing people for a common task if he wanted to make profitable use of his power.

However, the organizational structure of the new KWI für Physik did not correspond to this model. According to the founding proposal of February 1914, it was to be built on three bodies: a Kuratorium (administrative board) with one representative each from the Prussian Government, the KWG, and the Koppel-Stiftung; a "wissenschaftliches Komitee" (scientific committee) with a "permanent honorary secretary" (i.e., Einstein) and eight physicists elected for a period of three years—half of them from Berlin and half from elsewhere; and finally an "Arbeitsausschuss" (working committee). The intention was that the scientific committee should suggest to the Kuratorium research projects, the researchers to be selected, and the sums to be spent. The Kuratorium would then make the final decision concerning the funding of research projects and fellowships. The working committee would include all cooperating scientists funded by the institute and was to assemble at least once a year. The institute would not have a laboratory of its own.

> The *purpose of this institute* should be to form various groups of particularly competent researchers for the solution of important and urgent physical problems—either one after the other or simultaneously—in order to systematically bring the respective problems towards a possibly exhaustive solution through mathematical-physical examinations as well as by means of experimental investigations to be performed in the laboratories of the involved researchers.[55]

Clearly, the second "way" as suggested by Nernst during the meeting of 9 January 1914 had been taken. The model for this structure came from the Prussian Academy of Science. At the time, learned academies had become more or less "halls of fame" from which expert advice on the organization and support of research could be obtained and through which a certain amount of funds for research were still distributed. In fact, in the Berlin Academy, a "Geldverwendungs-Ausschuss" (appropriations committee) composed of five members of each of its two classes administered the capital endowed to the Academy by private donors. It backed applications for research projects by single persons as well as the joint efforts by groups of people, such as exploratory expeditions or the compilation of scientific dictionaries (Hartkopf and Wangermann 1991, 19, 24–27, 140–142). To us, it seems that the new KWI für Physik was conceived in analogy to the Geldverwendungs-Ausschuss and that its organization was modeled on an undertaking called Monumenta Germaniae Historica. This was a long-term editorial project for the publication of the sources of German history, promoted by the Academies of Berlin, Vienna, and Munich, and involving in a grand collaboration historians from all German-speaking countries (Grau 1975, 166). The Monumenta was given as an example in the proposal of February 1914:

---

[54]For a discussion of the "Harnack-Prinzip" and its concrete application, see Grau (1996), Laitko (1996), Vierhaus (1996). Adolf von Harnack was the founder and the first president of the Kaiser-Wilhelm-Gesellschaft.

[55]Kirsten and Treder (1979, 146, vol. 1). The emphasized part is underlined in original.

Such a cooperation of different scientists of a field is probably almost new in physics, but it is long since customary in other disciplines; in particular the committee for the "Monumenta Germaniae Historica" established by the Royal [Prussian] Academy of Science shall be remembered. (Kirsten and Treder 1979, 147, vol. 1)

The main structures of the Monumenta were a central directory board, composed of a minimum of eight members and a chairman, and a yearly plenary meeting of the board together with all leading editors. The Berlin members of the central directory board formed a steering committee in charge of everyday business between the plenary meetings. The general scientific and organizational questions were to be decided upon at the plenary meetings. The working agenda was then set by the central directory board and handed over to the scholars responsible for research and editorial work, who could hire junior collaborators themselves. They were all paid either a fixed salary or given funds for specific projects (Hartkopf and Wangermann 1991, 417–421). Thus, the organization of the KWI für Physik resembled that of the Monumenta Germaniae Historica.

Why was the organization chosen for Einstein's institute so different from that of the previously established KW Institutes, which were built around one or two creative scientists functioning as powerful directors?[56] We could speculate that the initiators were aware of Einstein's lack of experience in leading a large institute, but this would not explain the foundation of an institute with no laboratory at all. Of course, the proposers might just have been trying to obtain more funds for their own particular fields of research through the reputation of the young and promising director. After all, it must have been implicit from the beginning that they would become members of the leading boards of the new institute, as actually happened three years later. Through those committees they would then be able to influence the distribution of funds. This special structure must be seen against the background of the conflicts between the Prussian Academy, of which all the promoters, except Haber, were members, and the growing KWG. The Academy did not look favorably on the establishment of research institutes outside its control, but at the same time it was unable to adapt to the needs of advanced research (vom Brocke 1996; Grau 1993, 206–209). Since most of the members of the steering committees of the new KWI were to be de facto, if not de jure, members of the Academy, this particular structure would allow them and the Academy to command a greater budget and greater prestige. In a sense, the KWI für Physik was in fact an Academy Institute under the umbrella of the KWG. But also this explanation is unsatisfactory and only partial, since it would reduce the new foundation to mere maneuvering and ignore the long-running preliminaries and the scientific rationale cited in the founding documents.

Thus, on a less trivial note, we also should consider the assumption underlying the founding proposal, that in order to elaborate and pursue a research plan on the borderline between physics and physical chemistry, a collegial directorship would fare better than a single person. This interdisciplinary and discursive approach to

---

[56]The organizational difference from the other KWG institutes was even emphasized in the presentations of the KWI für Physik at the time of its foundation (see, e.g., Kaiser-Wilhelm-Gesellschaft 1918, 35).

research belonged to the constitutive tradition of the learned academies. Moreover, all undertakings of the Academy were organized in the form of working committees, like that of the Monumenta Germaniae Historica. Even undertakings in the natural sciences were not research institutes with their own laboratories, but "commissions" aimed at collecting and describing empirical data, like the "Catalogue of the Animal Kingdom" or the "History of the Firmament" (vom Brocke 1996, 9; Grau 1975, 154–156, 251–252). As for the internal scientific reasons for a collegial directorship, let us recall that, at least before Einstein's move to Berlin, a strong overlap existed between the research interests of the promoters and of Einstein, concentrating on the physical structure of matter. The hints coming in from so many subfields of physics that quantum phenomena would play an important role in the understanding of matter suggested the need for a cooperative effort, for an undertaking that could not possibly be led by just one creative scientist:

> Nowadays modern physics presents such important and vast problems, and in all probability this will be even more the case in future, that their ample treatment surpasses not only the forces of a single researcher but even those of a large physics institute.[57]

The scientific committee of the new KWI was meant to be the place to discuss these problems and to develop a research program to address them. This was the same expectation that Nernst, the original source of this kind of organizational solution, had nurtured towards the Solvay Congress of 1911.[58] Well aware of the contradictions between the new quantum theory and the classical molecular and kinetic theory of matter, he also believed, with a good dose of optimism, "that a personal discussion among researchers," (Kormos Barkan 1999, 189) theoreticians as well as experimentalists, would bring if not solutions at least indications as to the path to be followed. The letter of invitation to the congress, which had been drafted by Nernst, advocated "a conspicuous cooperative effort" in this sense between people working in different disciplines and using different approaches.

As for the leader of the cooperation, we have reason to believe that the promoters hoped that Einstein might provide inspiration for a breakthrough in the nebulous land of the mysterious quantum. It is too reductive to say that Einstein's "only task was to take part in the organization of research in the Academy as well as in the Kaiser Wilhelm Institute" (Frank 1949, 180), if we understand this only in the narrow sense of managing the staff's activities. Of course, to a certain extent, this was also part of Einstein's duties. Definitely, a managerial effort was required to start and follow through on a research project in which more than one scientist in possibly more than one location would take part. On the other hand, at the time, other important scientists were already known to be effective in this kind of work, while Einstein's organizational as well as experimental abilities were still untested. It seems to us that Einstein was chosen to become head of the institute above all because the new physics appearing on the horizon seemed to require a new step in the conceptual

---

[57]Foundation proposal, February 1914, in Kirsten and Treder (1979, 147, vol. 1).

[58]On Nernst's activities in the organization of scientific research and institutions, see Bartel and Huebener (2007, 206–229).

coordination of research. Because of his earlier work on the atomistic nature of matter and the quantum aspects of radiation, Einstein was believed to be capable of providing scientific leadership for the research carried out by other physicists.

Einstein himself was not entirely convinced of his ability to fulfill what was required of him and even worried about losing his job and income, if "sooner or later it turns out that I am not the right man for the direction of the institute and for the work connected with the fund to be established, so that I would feel obliged to resign my post."[59] He was evidently aware of the fact that becoming the Director of the KWI für Physik would involve scientific as well as administrative work and would not just mean obtaining an honorary title.

References

Bartel, Hans-Georg and Rudolf P. Huebener (2007). *Walther Nernst. Pioneer of Physics and Chemistry*. Singapore: World Scientific.

Bodenstein, Max (1942). "Walther Nernst, 25.6.1864–18.11.1941". In: *Berichte der Deutschen Chemischen Gesellschaft, Abteilung A* 75, pp. 79–104.

Bonhoeffer, Karl-Friedrich (1953). "Fritz Habers wissenschaftliches Werk". In: *Zeitschrift für Elektrochemie* 57, pp. 2–6.

Born, Max (1948–1949). "Max Karl Ernst Ludwig Planck". In: *Obituary Notices of Fellows of the Royal Society* 6, pp. 161–188.

Burchardt, Lothar (1975). *Wissenschaftspolitik im Wilhelminischen Deutschland. Vorgeschichte, Gründung und Aufbau der Kaiser-Wilhelm-Gesellschaft zur Förderung der Wissenschaften*. Göttingen: Vandenhoeck & Ruprecht.

Burchardt, Lothar (1988). "Naturwissenschaftliche Universitätslehrer im Kaiserreich". In: *Deutsche Hochschullehrer als Elite 1815–1945*. Ed. by Klaus Schwabe. Boppard am Rhein: Boldt, pp. 151–214.

Cahan, David (1989). *An Institute for an Empire. The Physikalisch-Technische Reichsanstalt 1871–1918*. Cambridge: Cambridge University Press.

Earman, John and Clark Glymour (1980). "The Gravitational Red Shift as a Test of General Relativity: History and Analysis". In: *Studies in History and Philosophy of Science* 11, pp. 175–214.

Einstein, Albert (1905). "Zur Elektrodynamik bewegter Körper". In: *Annalen der Physik* 17, pp. 891–921. Reprinted in (Einstein 1989, doc. 23).

Einstein, Albert (1907). "Die Plancksche Theorie der Strahlung und die Theorie der spezifischen Wärme". In: *Annalen der Physik* 22, pp. 180–190. Reprinted in (Einstein 1989, doc. 38).

Einstein, Albert (1909). "Über die Entwicklung unserer Anschauungen über das Wesen und die Konstitution der Strahlung". In: *Physikalische Zeitschrift* 10, pp. 817–825. Reprinted in (Einstein 1989, doc. 60).

Einstein, Albert (1912). "Thermodynamische Begründung des photochemischen Äquivalentgesetzes". In: *Annalen der Physik* 37, pp. 832–838. Reprinted in (Einstein 1995a, doc. 2).

Einstein, Albert (1979). *Autobiographical Notes*. Ed. by Paul Arthur Schilpp. La Salle and Chicago: Open Court Publishing Company. Translated and edited by Paul Arthur Schilpp.

Einstein, Albert (1993a). *The Collected Papers*. Vol. 3: *The Swiss Years: Writings, 1909–1911*. Ed. by Martin J. Klein et al. Princeton: Princeton University Press.

---

[59]Einstein to Planck, 7 July 1914 (Einstein 1998b, 40, doc. 18; Kirsten and Treder 1979, 104, vol. 1).

Einstein, Albert (1993b). *The Collected Papers*. Vol. 5: *The Swiss Years: Correspondence, 1902–1914*. Ed. by Martin J. Klein, Anne J. Kox, and Robert Schulmann. Princeton: Princeton University Press.

Einstein, Albert (1995a). *The Collected Papers*. Vol. 4: *The Swiss Years: Writings, 1912–1914*. Ed. by Martin J. Klein et al. Princeton: Princeton University Press.

Einstein, Albert (1995b). *The Collected Papers*. Vol. 5: *The Swiss Years: Correspondence, 1902–1914*. Princeton: Princeton University Press. English translation by Anna Beck.

Einstein, Albert (1998a). *The Collected Papers*. Vol. 8: *The Berlin Years: Correspondence, 1914–1918*. Princeton: Princeton University Press. English translation by Ann M. Hentschel.

Einstein, Albert (1998b). *The Collected Papers*. Vol. 8: *The Berlin Years: Correspondence, 1914–1918*. Ed. by Robert Schulmann et al. Princeton: Princeton University Press.

Einstein, Albert (2004a). *The Collected Papers*. Vol. 9: *The Berlin Years: Correspondence, January 1919–April 1920*. Princeton: Princeton University Press. English translation by Ann M. Hentschel.

Einstein, Albert (2004b). *The Collected Papers*. Vol. 9: *The Berlin Years: Correspondence, January 1919–April 1920*. Ed. by Diana Kormos Buchhwald et al. Princeton: Princeton University Press.

Einstein, Albert (2006). *The Collected Papers*. Vol. 10: *The Berlin Years: Correspondence, May–December 1920 and Supplementary Correspondence, 1909–1920*. Ed. by Diana Kormos Buchhwald et al. Princeton: Princeton University Press.

Einstein, Albert (2009). *The Collected Papers*. Vol. 12: *The Berlin Years: Correspondence, January–December 1921*. Ed. by Diana Kormos Buchhwald et al. Princeton: Princeton University Press.

Festschrift zur Feier des 500jährigen Bestehens der Universität Leipzig. 4. Band, 2. Teil: Die Institute und Seminare der philosophischen Fakultät an der Universität Leipzig. Die mathematisch-naturwissenschaftliche Sektion (1909). Leipzig: Hirzel.

Fölsing, Albrecht (1993). *Albert Einstein. Eine Biographie*. Frankfurt am Main: Suhrkamp.

Forbes, Eric G. (1972). "Freundlich, Erwin Finlay". In: *Dictionary of Scientific Biography*. Ed. by Charles C. Gillispie. Vol. 5. New York: Scribner, pp. 181–184.

Forman, Paul (1968). *The Environment and Practice of Atomic Physics in Weimar Germany*. Ann Arbor (Michigan): UMI.

Frank, Philipp (1949). *Einstein. Sein Leben und seine Zeit*. Munich: List.

Grau, Conrad (1975). *Die Berliner Akademie der Wissenschaften in der Zeit des Imperialismus*. Teil 1: *Von den neunziger Jahren des 19. Jahrhunderts bis zur Großen Sozialistischen Oktoberrevolution*. Berlin: Akademie-Verlag.

Grau, Conrad (1993). *Die Preußische Akademie der Wissenschaften zu Berlin. Eine deutsche Gelehrtengesellschaft in drei Jahrhunderten*. Berlin: Spektrum Akademischer Verlag.

Grau, Conrad (1996). "Genie und Kärrner—zu den geistesgeschichtlichen Wurzeln des Harnack-Prinzips in der Berliner Akademietradition". In: *Die Kaiser-Wilhelm-/Max-Planck-Gesellschaft und ihre Institute*. Ed. by Bernhard vom Brocke and Hubert Laitko. Berlin: de Gruyter, pp. 139–144.

Hartkopf, Werner and Gert Wangermann, eds. (1991). *Dokumente zur Geschichte der Berliner Akademie der Wissenschaften von 1700 bis 1990*. Berlin: Spektrum Akademischer Verlag.

Heilbron, John L. (1986). *The Dilemma of an Upright Man: Max Planck as Spokesman for German Science*. Berkeley (Cal.): University of California Press.

Hentschel, Klaus (1992). "Einstein's Attitude towards Experiments: Testing Relativity Theory 1907–1927". In: *Studies in History and Philosophy of Science* 23, pp. 593–624.

Hentschel, Klaus (1994). "Erwin Finlay Freundlich and Testing Einstein's Theory of Relativity". In: *Archive for History of Exact Sciences* 47, pp. 143–201.

Hentschel, Klaus (1998). *Zum Zusammenspiel von Instrument, Experiment und Theorie. Rotverschiebung im Sonnenspektrum und verwandte spektrale Verschiebungseffekte von 1880 bis 1960*. Hamburg: Kovac.

Hermann, Armin (1978). "Geschichte der physikalischen Institute im Deutschland des 19. Jahrhunderts". In: *Das Forschungsinstitut. Formen der Institutionalisierung von Wissenschaft*. Ed. by Erwin K. Scheuch and Heine von Aleman. Erlangen: Deutsche Gesellschaft für zeitgeschichtliche Fragen, pp. 95–118.

Hermann, Armin (1994). *Einstein. Der Weltweise und sein Jahrhundert*. Munich: Piper.

Hiebert, Erwin N. (1978). "Nernst, Hermann Walther". In: *Dictionary of Scientific Biography*. Ed. by Charles C. Gillispie. Vol. 15, Supplement I. New York: Scribner, pp. 432–453.

Hoffmann, Dieter (1984). "Max Planck als akademischer Lehrer". In: *Berliner Wissenschaftshistorische Kolloquien VIII: Die Entwicklung der Physik in Berlin*. Berlin: Akademie der Wissenschaften der DDR. Institut für Theorie, Geschichte und Organisation der Wissenschaft, pp. 55–71.

Jastrow, Ignaz (1930). "Kollegiengelder und Gebühren". In: *Das akademische Deutschland. Band III: Die deutschen Hochschulen in ihren Beziehungen zur Gegenwartskultur*. Ed. by Michael Doeberl et al. Berlin: Weller, pp. 277–284.

Jungnickel, Christa and Russell McCormmach (1986). *Intellectual Mastery of Nature. Theoretical Physics from Ohm to Einstein*. Chicago: The University of Chicago Press.

Kaiser-Wilhelm-Gesellschaft (1918). "Die Institute und Unternehmungen der Kaiser-Wilhelm-Gesellschaft zur Förderung der Wissenschaften". In: *Die Naturwissenschafen* 6, pp. 34–35.

Kangro, Hans (1975). "Rubens, Heinrich". In: *Dictionary of Scientific Biography*. Ed. by Charles C. Gillispie. Vol. 11. New York: Scribner, pp. 581–585.

Kant, Horst (1996). "Albert Einstein, Max von Laue, Peter Debye und das Kaiser-Wilhelm-Institut für Physik in Berlin (1917–1939)". In: *Die Kaiser-Wilhelm-/Max-Planck-Gesellschaft und ihre Institute*. Ed. by Bernhard vom Brocke and Hubert Laitko. Berlin: de Gruyter, pp. 227–243.

Kant, Horst and Dieter Hoffmann (1981). "Die Physik in Berlin von der Universitätsgründung bis zur Jahrhundertwende—Institutionalisierung, Hauptarbeitsgebiete, Wechselwirkung mit der Industrie". In: *Berliner Wissenschaftshistorische Kolloquien III: Die Entwicklung Berlins als Wissenschaftszentrum (1870–1930)*. Berlin: Akademie der Wissenschaften der DDR. Institut für Theorie, Geschichte und Organisation der Wissenschaft, pp. 129–175.

Kirsten, Christa and Hans-Jürgen Treder, eds. (1979). *Albert Einstein in Berlin 1913–1933*. Berlin: Akademie-Verlag.

Kormos Barkan, Diana (1999). *Walther Nernst and the Transition to Modern Physical Science*. Cambridge: Cambridge University Press.

Kox, Anne J. (1995). "Einstein, Specific Heats, and Residual Rays: The History of a Retracted Paper". In: *No Truth Except in the Details*. Ed. by Anne J. Kox and Daniel Siegel. Dordrecht: Kluwer Academic, pp. 245–257.

Kragh, Helge (1999). *Quantum Generations. A History of Physics in the Twentieth Century*. Princeton: Princeton University Press.

Laitko, Hubert (1996). "Persönlichkeitszentrierte Forschungsorganisation als Leitgedanke der Kaiser-Wilhelm-Gesellschaft: Reichweite und Grenzen, Ideal und Wirklichkeit". In: *Die Kaiser-Wilhelm-/Max-Planck-Gesellschaft und ihre Institute*. Ed. by Bernhard vom Brocke and Hubert Laitko. Berlin: De Gruyter, pp. 583–632.

Langevin, Paul and Maurice de Broglie, eds. (1912). *La théorie du rayonnement et les quanta. Rapports et discussions de la Réunion tenue à Bruxelles, du 30 octobre au 3 novembre 1911, sous les Auspices de M. E. Solvay*. Paris: Gauthier-Villars.

Laue, Max (1911). *Das Relativitätsprinzip*. Braunschweig: Vieweg.

Mehra, Jagdisch and Helmut Rechenberg (1982). *The Historical Development of Quantum Theory*. Vol. 1: *The Quantum Theory of Planck, Einstein, Bohr and Sommerfeld: Its Foundation and the Rise of Its Difficulties 1900–1925*. New York: Springer.

Nernst, Walther (1913). *Theoretische Chemie vom Standpunkte der Avogadroschen Regel und der Thermodynamik*. 7th ed. Stuttgart: Enke.

Pais, Abraham (1982). *"Subtle is the Lord ..." The Science and the Life of Albert Einstein*. Oxford: Oxford University Press.

Planck, Max (1913). *Vorlesungen über die Theorie der Wärmestrahlung*. 2nd ed. Leipzig: Barth.

Planck, Max (1914). "Die Gesetze der Wärmestrahlung und die Hypothese der elementaren Wirkungsquanten". In: *Die Theorie der Strahlung und der Quanten. Verhandlungen auf einer von E. Solvay einberufenen Zusammenkunft (30. Oktober bis 3. November 1911)*. Ed. by

Arnold Eucken. Abhandlungen der Deutschen Bunsen-Gesellschaft für angewandte physikalische Chemie, 7. Halle an der Saale: Knapp, pp. 77–108.

Pyenson, Lewis (1985). *The Young Einstein. The Advent of Relativity*. Bristol: Hilger.

Ramsauer, Carl (1913). "Das physikalisch-radiologische Institut der Universität Heidelberg". In: *Frankurter Zeitung* 145, 25 July 1913. Reprinted in (Auer 1984, 57–62).

Ramser, Hans (1976). "Warburg, Emil Gabriel". In: *Dictionary of Scientific Biography*. Ed. by Charles C. Gillispie. Vol. 14. New York: Scribner, pp. 170–172.

Riecke, Eduard (1905). "Das neue physikalische Institut der Universität Göttingen". In: *Physikalische Zeitschrift* 6, pp. 881–892.

Riese, Reinhard (1977). *Die Hochschule auf dem Wege zum wissenschaftlichen Großbetrieb. Die Universität Heidelberg und das badische Hochschulwesen 1860–1914*. Stuttgart: Klett.

Ritter, Gerhard A. (1992). *Großforschung und Staat in Deutschland. Ein historischer Überblick*. Munich: Beck.

Scheuch, Erwin K. and Heine von Aleman, eds. (1978). *Das Forschungsinstitut. Formen der Institutionalisierung von Wissenschaft*. Erlangen: Deutsche Gesellschaft für zeitgeschichtliche Fragen.

Schulmann Robert (1995) "From Periphery to Center: Einstein's Path from Bern to Berlin (1902–1914)". In: *No Truth Except in the Details*. Ed. by Anne J. Kox and Daniel Siegel. Dordrecht: Kluwer Academic, pp. 259–271.

Seelig, Carl (1960). *Albert Einstein. Leben und Werk eines Genies unserer Zeit*. Zurich: Europa Verlag.

Stobbe, Hans (1936–1940). *J. C. Poggendorffs biographisch-literarisches Handwörterbuch*. Band VI: *1923 bis 1931*. Berlin: Verlag Chemie.

Stoltzenberg, Dietrich (1994). *Fritz Haber: Chemiker, Nobelpreisträger, Deutscher, Jude; eine Biographie*. Weinheim: VCH.

Szöllösi-Janze, Margit (1998). *Fritz Haber 1868–1934. Eine Biographie*. Munich: Beck.

Vierhaus, Rudolf (1996). "Bemerkungen zum sogennanten Harnack-Prinzip. Mythos und Realität". In: *Die Kaiser-Wilhelm-/Max-Planck-Gesellschaft und ihre Institute*. Ed. by Bernhard vom Brocke and Hubert Laitko. Berlin: de Gruyter, pp. 129–138.

Vierhaus, Rudolf and Bernhard vom Brocke, eds. (1990). *Forschung im Spannungsfeld von Politik und Gesellschaft. Geschichte und Struktur der Kaiser-Wilhelm-/Max Planck-Gesellschaft*. Stuttgart: Deutsche Verlagsanstalt.

vom Brocke, Bernhard (1990). "Die Kaiser-Wilhelm-Gesellschaft im Kaiserreich. Vorgeschichte, Gründung und Entwicklung bis zum Ausbruch des Ersten Weltkriegs". In: *Forschung im Spannungsfeld von Politik und Gesellschaft. Geschichte und Struktur der Kaiser-Wilhelm-/Max-Planck-Gesellschaft*. Ed. by Rudolf Vierhaus and Bernhard vom Brocke. Stuttgart: Deutsche Verlagsanstalt, pp. 17–162.

vom Brocke, Bernhard (1996). "Die Kaiser-Wilhelm-/Max-Planck-Gesellschaft und ihre Institute zwischen Universität und Akademie. Strukturprobleme und Historiographie". In: *Die Kaiser-Wilhelm-/Max-Planck-Gesellschaft und ihre Institute*. Ed. by Bernhard vom Brocke and Hubert Laitko. Berlin: de Gruyter, pp. 1–32.

vom Brocke, Bernhard and Hubert Laitko, eds. (1996). *Die Kaiser Wilhelm-/Max Planck-Gesellschaft und ihre Institute*. Berlin: de Gruyter.

von Staa, Meinhard (1930). "Aufbau und Bedeutung der deutschen Universitätsinstitute und Seminare". In: *Das akademische Deutschland*. Band III: *Die deutschen Hochschulen in ihren Beziehungen zur Gegenwartskultur*. Ed. by Michael Doeberl et al. Berlin: Weller, pp. 263–276.

Weinmeister, Paul (1925–1926). *J. C. Poggendorffs biographisch-literarisches Handwörterbuch*. Band V: *1904 bis 1922*. Leipzig: Verlag Chemie.

Wendel, Günter (1975). *Die Kaiser-Wilhelm-Gesellschaft 1911–1914. Zur Anatomie einer imperialistischen Forschungsgesellschaft*. Berlin: Akademie-Verlag.

Wendel, Günter (1984). "Die Berliner Institute der Kaiser-Wilhelm-Gesellschaft und ihr Platz im System der Wissenschaftspolitik des imperialistischen Deutschland in der Zeit bis 1933". In: *Berliner Wissenschaftshistorische Kolloquien X: Zur Geschichte der Deutschen Staatsbibliothek sowie zu Tendenzen der Wissenschaftspolitik im imperialistischen Deutschland*. Berlin: Akademie der Wissenschaften der DDR. Institut für Theorie, Geschichte und Organisation der Wissenschaft, pp. 27–69.

Wiener, Otto (1906). "Das neue physikalische Institut der Universität Leipzig und Geschichtliches". In: *Physikalische Zeitschrift* 7, pp. 1–14.

Zaunick, Rudolf and Hans Salié (1956–1962). *J. C. Poggendorffs biographisch-literarisches Handwörterbuch der exakten Naturwissenschaften*. Band VIIa: *Berichtsjahre 1932 bis 1953*. Berlin: Akademie-Verlag.

Chapter 2
The Foundation of the KWI für Physik

Abstract This chapter addresses the failed attempt to open the Kaiser-Wilhelm-Institut für physikalische Forschung (KWI für Physik) in Berlin in 1914, Einstein's early years in Berlin, and the successful foundation of the institute in 1917. The deviation of the organizational structure of this new institute from existing Kaiser Wilhelm Institutes is also discussed as being due to changes in the conceptual coordination of research, which began to be steered by the scientific community.

Keywords Albert Einstein · Fritz Haber · Adolf von Harnack · Philipp Lenard · Walter Nernst · Max Planck · Heinrich Rubens · Prussian Ministry · Koppel-Stiftung

2.1   The Preliminaries. A First Unsuccessful Attempt to Establish the Institute

As early as 1906, the experimental physicist and Nobel Prize laureate Philipp Lenard had submitted to the Prussian Ministry of Education a memorandum concerning the establishment of an "Institute for Physical Research."[1] Lenard argued that pure physics research was necessary for technological development. In particular, he mentioned the development of the electrotechnical industry with its applications in transport and communications systems and illumination techniques. But, in his opinion, the way in which research was organized and carried out in the university institutes and the Physikalisch-Technische Reichsanstalt was inadequate. In order to be more effective, "a number of able personalities ought to be joined in a specially established institute with the necessary financial means, an Institute for Physical Research." "The task of the institute should be the productive cultivation of physical research through

---

[1]Burchardt (1975, 22), Kant (1987, 129–130). For the history of the KWI für Physik, see Heisenberg (1971), Kant (1987, 1989, 1992, 1993, 1996), Schlüter (1994, 1995). The bulk of the documents relating to the KWI für Physik are kept at the Archiv zur Geschichte der Max-Planck-Gesellschaft [hereafter AMPG] in Berlin, I. Abt., Rep. 1 A, Nr. 1649–1671; the documents of the KWI für Physik itself are in AMPG, I. Abt., Rep. 34.

© The Author(s), under exclusive license to Springer Nature Switzerland AG 2020
H. Goenner and G. Castagnetti, *Establishing Quantum Physics in Berlin*,
SpringerBriefs in History of Science and Technology,
https://doi.org/10.1007/978-3-030-63122-2_2

continuous experimental investigations."[2] Nowhere in Lenard's memorandum were the most recent developments in radiation physics or radioactivity mentioned.

Lenard's memorandum was not acted upon in Berlin.[3] Nevertheless, discussions continued about the necessity of a specialized institute for advanced research in physics, unrestrained by the universities' institutional duties. A new suggestion was made in 1908 by Nernst, who proposed the foundation of a "Research Institute for Radioactivity and Electronics" aimed "at the latest developments on the borderland between physics and chemistry" (Kant 1987, 130; Burchardt 1975, 27). But the plan became more concrete only after the foundation of the KWG in 1911. In a meeting on 5 February 1912, Schmidt-Ott and Krüss, as representatives of the Prussian Ministry of Education, discussed with Nernst, Haber, Koppel, Emil Fischer,[4] and the President of the KWG, Harnack, the foundation of a "Radiological Institute" to be built around Rutherford, in whom Fischer in particular was interested. However, neither Haber and Nernst nor Harnack were enthusiastic about this proposal; a "general physics" institute was more to their liking and Wien was mentioned as its director. Nernst expressed a preference for Wien rather than Lenard because of Wien's broader interests. On this occasion, Einstein had not yet been mentioned.[5] Only in 1913 did the idea emerge to combine the plan for a new physics institute with the aim of recruiting Einstein to Berlin. Nevertheless, alternatives were kept open up until the meeting of 9 January 1914. On that occasion, as we have seen, Nernst again suggested the establishment of an "Institute for Radiology" with Geiger as a possible director or, alternatively, the constitution of a scientific committee with Einstein among its members.[6] This was the solution that was finally adopted.

As mentioned earlier, on 4 February 1914 Nernst and his colleagues sent a proposal to the Prussian Ministry of Education, to the KWG, and to the Koppel-Stiftung concerning the establishment of a KWI "für physikalische Forschung" (for physical research). In addition to an exposition of the aims and organization of the new institute, the proposal also gave a first estimate of 75,000 marks for its annual budget (Kirsten and Treder 1979, 146, vol. 1). In the following months, the different aspects of the plan were worked out in coordination with the proposers and representatives of the institutions concerned. A few weeks later, Harnack communicated to the Koppel-Stiftung his preliminary approval for the project and presented a financial plan for the first ten years of the institute. It amounted to splitting the financial support into three equal parts of 25,000 marks to be provided by the KWG, by the Koppel-Stiftung, and

---

[2]"Denkschrift und Entwurf zu einem deutschen Institut für physikalischen Forschung," December 1906, GStA, I. HA, Rep. 92 Althoff, A I, Nr. 123, pp. 66–76. The sentences quoted are on pp. 5 and 12 respectively.

[3]An institute in accordance with Lenard's suggestions was opened in 1913 in Heidelberg (Auer 1984).

[4]Fischer was professor of chemistry at the University of Berlin and member of the steering board of the KWG.

[5]Minutes of the meetings, GStA, I. HA, Rep. 76 76 V c, Sekt. 2, Tit. 23, Litt. A, Nr. 116, pp. 16–17.

[6]Minutes of the meetings, GStA, I. HA, Rep. 76 76 V c, Sekt. 2, Tit. 23, Litt. A, Nr. 116, pp. 18 r–v; Wendel (1975, 197).

by the Prussian government. Furthermore, Harnack welcomed Koppel's intention to take over the costs for the construction of the institute's building.[7]

On 2 March 1914, Nernst provided Krüss at the Ministry of Education with an additional explanatory exposé as an aid to negotiations with the Ministry of Finance.[8] The main argument for the new foundation was that the institute would be able to better allocate the resources and concentrate efforts on tackling "bigger experimental works" which were beyond the reach of a single university institute. Furthermore, as Lenard had done, Nernst pointed out the benefits of the advancement of physics research for German industry, namely for the improvement of precision mechanics and the production of X-ray instruments. In passing, the proposed institute was called "Kaiser-Wilhelm-Institut für theoretische Physik" (for theoretical physics) instead of "für physikalische Forschung" as in the February proposal.

On another front, Harnack too explained the main points of the foundation plan to Gustav Krupp von Bohlen und Halbach, the chairman of the famous steel company Krupp and vice president of the KWG.[9] Harnack stressed that the new institute would not itself perform the "research on theoretical physics now essential to this science," but entrust it to the existing university institutes. It would buy the necessary instruments and lend them to the laboratories for as long as they were needed. The Koppel-Stiftung would also donate a "small, but beautiful building" to the KWG as the main office for the steering committees of the new institute, and as a storehouse for its instruments. This building would also serve as a meeting point for the scientists of the other KW Institutes situated nearby in Berlin-Dahlem, and would feature a large hall for the official gatherings of the KWG.[10] In light of this generous offer, Harnack concluded that the KWG should take up the project and delay the establishment of an institute for history.

On 21 March 1914, the administrative committee and the "Senat" (steering board) of the KWG decided to accept the offer of the Koppel-Stiftung to contribute a building and one third of the running expenses for ten years and thus to establish the new institute. The KWG would take over another third of the costs while the State of Prussia would finance the remaining third.[11] An architect was contacted soon thereafter for the construction of a building on the grounds of the KWI für Chemi.e.[12] The steering board of the Koppel-Stiftung first discussed the project officially on 2 April 1914. In addition to the annual contribution for ten years, it decided to bestow

---

[7] Von Harnack to Koppel-Stiftung, 26 February 1914, GStA, I. HA, Rep. 92 Schmidt-Ott, B LXXVI, Bd. 4, pp. 17–19; Burchardt (1975, 118–119), Wendel (1975, 199).

[8] "Im nachfolgenden beehre ich mich ...," 2 March 1914, SBB, Acta Preussische Staatsbibliothek, Generaldirektion, Kaiser-Wilhelm-Institute XXVI. The typewritten document is not signed, but the corrections are by Nernst's hand.

[9] Von Harnack to Krupp von Bohlen und Halbach, 4 March 1914, Historisches Archiv Krupp, Essen, FAH 4 E 277, pp. 20–21.

[10] This idea was realized in 1929 with the opening of the "Harnack-Haus" (Henning 1996).

[11] Minutes of the meeting, GStA, I. HA, Rep. 89 (2.2.1.), Nr. 21289, pp. 180–184; Burchardt (1975, 118).

[12] Carl Duisberg to *Verwaltungsrat* of the KWI for Chemistry, 1 April 1914, GStA, I. HA, Rep. 92 Schmidt-Ott, C, Bd. 84, pp. 15–16; Kirsten and Treder (1979, 67, vol. 2).

a total of 625,000 marks upon the institute at the end of this period, on the condition that the Prussian government and the KWG would then guarantee their support for an unlimited period of time.[13]

Towards the end of April 1914, Haber gave Krüss another memorandum concerning the KWI "für theoretische Physik," most probably as an aid for the formulation of the proposal that the Ministry of Education had to submit to the Ministry of Finance.[14] With arguments similar to those of Nernst, Haber explained that this kind of institute would allow a better distribution of the available funds. In particular, the document pointed out the theoretical crisis affecting physics as well as the urgent need for new experimental investigations in order to obtain "clear physical knowledge." A more coherent theoretical system of the "foundations" of physics would also support the development of technology as well as improve the use of X-rays in medicine. The emphasis put on the need for progress in theoretical physics explains the institute's name, adopted by both Haber and Nernst, and reveals an implicit consent among the promoters. To us, the fact that the name of the new institute had become "for theoretical physics" instead of the previously suggested "for physical research," which in Lenard's original use meant experimental physics only, is a further indication of the prominent role ascribed by the Berlin chemistry and physics professors to the theoretician Einstein.[15]

On 5 June 1914 the KWG and the Koppel-Stiftung submitted the official joint proposal for the establishment of a KWI "für theoretische Physik" to the Prussian Ministry of Education. The rationale for the new institute was taken from the February memorandum, stressing that "in the field of theoretical physics important and basic questions are awaiting a solution." The financial plan was the one agreed upon by the KWG and the Koppel-Stiftung. Koppel reiterated his intention to personally take over the costs for the construction of the building.[16] Finally, a month later, the Prussian Minister of Education submitted the project to the Prussian Ministry of Finance.[17] First of all, it was pointed out that the new institute under Einstein's directorship would carry out almost no research work of its own but rather distribute its funds to existing laboratories. Haber's arguments were then taken up verbatim, starting with a short history of theoretical physics of the last decade.

---

[13]Minutes of the meeting, Bundesarchiv Berlin, REM, Nr. 1153, pp. 277–278; Wendel (1975, 199). In order to finance the KWI für Physik, the Foundation gave up the plan to establish a home for female workers and Koppel personally added 85,000 marks to the fund (Report of the Koppel-Stiftung to Kaiser Wilhelm II, 6 June 1914, GStA, I. HA, Rep. 89 (2.2.1.), Nr. 20008, pp. 188–194).

[14]"Gegen Ende des vorigen Jahrhunderts …" SBB, Acta Preussische Staatsbibliothek, Generaldirektion, Kaiser-Wilhelm-Institute XXVI. The document was received by Krüss on 24 April.

[15]However, the institute opened in October 1917 as "Kaiser-Wilhelm-Institut für physikalische Forschung" (see Sect. 2.3). The name "Kaiser-Wilhelm-Institut für Physik" was finally suggested by Harnack, probably in analogy with the more general name of the other KW Institutes, and approved by Kaiser Wilhelm II in December 1917 (GStA, I. HA, Rep. 89 (2.2.1.), 21306, pp. 37–40).

[16]Von Harnack and Koppel to Prussian Minister of Education, GStA, I. HA, Rep. 76 76 V c, Sekt. 2, Tit. 23, Litt. A, Nr. 116, pp. 6–8.

[17]Prussian Minister of Education to Prussian Minister of Finance, 2 July 1914, GStA, I. HA, Rep. 76 76 V c, Sekt. 2, Tit. 23, Litt. A, Nr. 116, pp. 21–24; Wendel (1975, 199–200).

> About ten years ago it was discovered that the until then apparently secure basic principles of theoretical physics were in deep contradiction with new experimentally acquired results. This discovery has been supported by the fundamental research of the German scientist Planck, who established that the observations in the field of heat radiation were accountable only with the help of a basic idea totally at odds with the previous theory of nature. [...] Einstein then demonstrated that this basic idea has far-reaching consequences in the field of the molecular properties of matter and Nernst could finally bring forth the evidence confirming those consequences, which would never have been believable before.

After the scientific explanations the Minister upped the ante with a hint at the positive effects that basic research would have on technical progress and with an appeal to national pride. More funds were needed in order to develop truly German discoveries and thus catch up with England and France in the scientific race. These arguments, however, did not convince the Ministry of Finance. On 31 July 1914, one day before the First World War began, it turned down the plan with formal and fiscal reasoning. Even if the need for additional funds for physical research had to be acknowledged,

> [...] the obvious course of action would be to hand out the state funds directly to the concerned state and university institutes and to avoid making the totally unnecessary and in principle contestable detour through funding the Kaiser-Wilhelm-Gesellschaft. The idea of a systematic and uniform approach toward the envisaged aims could also fully be achieved in this way.[18]

Nernst had foreseen this argument in his note to Krüss of 2 March 1914. According to him, the obvious solution, that is, an increase in the budgets of the university institutes, would not work. First of all, the money for this would be lacking. The KWG was able to collect funds from private donors, thus considerably surpassing the means the state could allocate for pure research. Second, the state was also supporting institutes in which research was stagnating. In contrast, the KWI would be able to invest the money "where real personalities are working who would make good use of it."[19] The same argument had been used in the joint proposal of the Berlin physicists of February 1914: "[O]bviously, only such physical laboratories which house lively scientific activities would enjoy the benefits of the new institution since only industrious and successful researchers will be involved in collaboration" (Kirsten and Treder 1979, 147, vol. 1).

In fact, the Minister of Finance loathed the idea of having another KWI. His refusal is a fine example of misunderstanding and mistrust of goings-on among scholars. At the same time as he denied funding a new research institute in Berlin, he took the occasion to clear away another physics-related initiative in Göttingen:

---

[18]Prussian Minister of Finance to Prussian Minister of Education, 31 July 1914, GStA, I. HA, Rep. 76 V c, Sekt. 2, Tit. 23, Litt. A, Nr. 116, pp. 25–27; Wendel (1975, 201).

[19]"Im nachfolgenden beehre ich mich ..." 2 March 1914, SBB, Acta Preussische Staatsbibliothek, Generaldirektion, Kaiser-Wilhelm-Institute XXVI.

By the way I remark that Professor Hilpert [sic] from Göttingen made mention in my Ministry of an initiative seemingly in the same field [as the Berlin initiative] by the local Academy of Sciences, for which funds were also to be requested. I assume that these plans will now be taken up in general by the Association for Theoretical Physics [Vereinigung für theoretische Physik] and that thereby the particular project for Göttingen will be superseded.[20]

In contrast, the Minister hinted at his eventual willingness to supply further financial support through the existing structures, though without making any commitment.

Although nothing in the letter indicates it, the Ministry, in view of mounting tensions in international politics, might just have been playing it safe, namely by not spending money. In any case, on 12 August 1914, following the outbreak of the First World War, the administrative committee of the KWG decided to postpone several projects, among them the establishment of the "KWI für Physikalische Forschung."[21] In October, the KWG and the Koppel-Stiftung both informed the Ministry of Education that the new institute would not be opened for the time being.[22]

2.2 Einstein's First Years in Berlin

Einstein's first six months in Berlin were clouded by two disasters.

In June 1914, Einstein's marriage was definitely over. His wife Mileva left Berlin and returned to Zurich with their two sons (Fölsing 1993, 383; Hermann 1994, 211). A few weeks later, the First World War began. Einstein reacted to these private and public turmoils by taking refuge in his scientific work and in the relationship with his cousin Elsa. His letters from that time contain numerous expressions of his satisfaction with the working conditions and the Berlin environment. Of course, he hated the chauvinistic attitude of many of his colleagues and felt isolated by them and thus associated with pacifist circles (Nathan and Norden 1968, 1–2; Fölsing 1993, 393–395; Goenner and Castagnetti 1996).

Although Einstein still continued to speak about Nernst's heat theorem and a thermodynamic derivation of Planck's radiation formula in July 1914 (Einstein 1914a), after his arrival in Berlin his scientific work was at first almost unilaterally directed toward the completion of general relativity. Six of his nine research papers written from 1914 to 1915 in Berlin deal with this topic (Einstein 1914b, c, 1915a, b, c, d). In autumn 1915, Einstein finally obtained the famous field equations of general relativity after months of pursuing an arduous and convoluted path, thus revolutionizing thinking about space, time, and gravitation.[23] During this time he also tried to discern possible consequences of general relativity and to find a way to test it empirically

---

[20]Prussian Minister of Finance to Prussian Minister of Education, 31 July 1914, GStA, I. HA, Rep. 76 V c, Sekt. 2, Tit. 23, Litt. A, Nr. 116, pp. 25–27. "Hilpert," of course, is David Hilbert. The loose group of Berlin initiators has become a non-existent "association."

[21]Minutes of the meeting, GStA, I. HA, Rep. 89 (2.2.1.), Nr. 21289, p. 232; Wendel (1975, 350).

[22]Von Harnack and Koppel to Prussian Minister of Education, 3 October 1914, GStA, I. HA, Rep. 76 76 V c, Sekt. 2, Tit. 23, Litt. A, Nr. 116, p. 28.

[23]In fact, as we now know, he returned to them (Renn and Sauer 2007; Janssen and Renn 2007).

(Hentschel 1992a, b, 1994). At the same time, Einstein worked as a guest scientist in the laboratory of the Physikalisch-Technische Reichsanstalt, where, together with Wander J. de Haas, he performed an experiment on magnetism and discovered what is now called the Einstein-de Haas effect (Hoffmann 1980, 91–92; Einstein and de Haas 1915).

Only in 1916 did Einstein return to problems in the quantum theory of radiation, on which some of his Berlin colleagues had been busily at work. The general theoretical context was as follows[24]: Since 1913 Bohr's atomic model had been able to account, to some degree, for the observations of the line spectrum of hydrogen. In 1915 Sommerfeld, in Munich, improved the model by exchanging Bohr's special circular orbits of the electron around the hydrogen nucleus with the more general elliptical orbits and by taking into account the velocity dependence of the electron's mass. As a result, the fine structure of spectra, that is, line multiplets, could be explained. In Berlin, Planck proposed a quantum condition in phase space for molecules with several degrees of freedom. Similarly, the astronomer Schwarzschild tried to find the proper coordinates for which the quantization conditions applied to systems with several degrees of freedom and looked at the Stark effect, that is, the splitting of spectral lines in an electrical field. Rubens and his assistant Gerhard Hettner investigated the long-wavelength side of the spectrum of hydrogen vapor and tried to interpret the measurement's results by means of quantum theory (Rubens and Hettner 1916). Warburg continued his experiments concerning photochemistry in gases (Warburg 1916). Einstein worked on spontaneous and induced emission and absorption, deriving afresh Planck's radiation formula by means of these concepts (Einstein 1916). In 1917, he took up what he called "the Sommerfeld-Epstein formulation of quantum theory"[25] and dealt with the question of the correct coordinates for the action integrals to be quantized; he gave a quantization condition independent of the choice of the coordinates (Einstein 1917b).

Since autumn 1916 Einstein's thinking had also been occupied by questions concerning the large-scale distribution of stars, or what is now called cosmology. He had extended discussions and correspondence on this subject with Willem de Sitter, Besso, and others.[26] In February 1917 he then published his seminal paper introducing the cosmological constant and containing the first cosmological model, later named the Einstein cosmos (Einstein 1917a).

Meanwhile, Einstein must have given up hope that he would get an institute of his own. Although a university professor and Academy member, he had no institutional power whatsoever and no place to work apart from his own flat. He apparently did not use the office at his disposal in Haber's institute. Einstein's efforts to test general relativity with the help of Freundlich were hampered by Freundlich's director at the

---

[24]For a detailed account and the bibliography, see Kragh (1999), Mehra and Rechenberg (1982); see also Jungnickel and McCormmach (1986, vol. 2).

[25]Einstein to Besso, 29 April 1917, in Einstein (1998, 442, doc. 331).

[26]See the editorial comment "The Einstein-de Sitter-Weyl-Klein-Debate" in Einstein (1998, 351–357), and the documents referred to therein. See also Kerszberg (1989).

Sternwarte—and Einstein's colleague in the Academy—Hermann Struve (Hentschel 1992a, 52–55). German astronomers at that time were almost all anti-relativists. Since 1915 Einstein had tried unsuccessfully to have Freundlich released from his duties, so that he could devote himself to the observation of light deflections produced by gravitational fields, but Struve would not allow Freundlich to work freely. In November 1915 Einstein complained to Sommerfeld that "the intrigues of miserable people" were hindering Freundlich in this work.[27] Only in December 1916 was Einstein appointed to the scientific steering board of the Physikalisch-Technische Reichsanstalt at the suggestion of its President, Warburg (Hoffmann 1980, 92–93). This was only a partial remedy to Einstein's isolation, of which his colleagues were probably aware.

2.3 The Opening of the KWI für Physik

The situation changed in January 1917, as the Berlin industrialist Franz Stock joined the KWG by donating the considerable entry fee of 540,000 marks; 500,000 marks were in the form of war bonds, the return of which was earmarked for physics research.[28] The KWG took this opportunity to reconsider opening the institute promised to Einstein. On 6 July 1917, after an exchange of letters between Harnack, the Koppel-Stiftung, and the Ministry for Education, and after a discussion with the "Berlin physicists" had taken place on 26 June, the Senat of the KWG decided to establish a KWI for physical research as of 1 October 1917.[29] Its yearly budget would amount, as foreseen, to 75,000 marks,[30] the KWG now contributing one-third for an unlimited period of time and also taking over for ten years the share originally to be provided by the Prussian government. As promised in 1914, the Koppel-Stiftung would contribute the remaining 25,000 marks for ten years, so that for the time being the new institute would be financed entirely by private donors. Einstein would be made director with an annual salary of 5,000 marks. The construction of a building for the new institute was no longer mentioned (Table 2.1).

In comparison with some of the other Kaiser Wilhelm Institutes already in existence, the budget of the KWI für Physik was not very large. Haber's KWI für physikalische Chemie und Elektrochemie had an annual budget of 100,000 marks and received from the Koppel-Stiftung a starting endowment of more than one million marks. The KWI für Chemie had a yearly budget of 133,700 marks and a starting fund of 1,100,000 marks. Also the KWI für Biologie (140,000 marks), für Kohlen-

---

[27]Einstein to Sommerfeld, 28 November 1915, in Einstein (1998, 208, doc. 153).

[28]Von Harnack to the Senat members, 18 January 1917, GStA, I. HA, Rep. 76 76 V c, Sekt. 2, Tit. 23, Litt. A, Nr. 116, p. 33.

[29]Minutes of the Senat meeting, 6 July 1917, AMPG, I. Abt., Rep. 1 A, Nr. 1656, p. 1. See also von Harnack to Schmidt-Ott, 15 February 1917, GStA, I. HA, Rep. 92 Schmidt-Ott, B LXXVI, Bd. 4, p. 20; von Harnack to Schmidt-Ott, 12 September 1917, in Kirsten and Treder (1979, 148–149, vol. 1), Wendel (1975, 223, 358–359), Burchardt (1990, 177).

[30]"[A]bout 15,000 USD at the time" (Forman 1968, 259).

Table 2.1 Promised annual contributions to the KWI für Physik at its start

Kaiser-Wilhelm-Gesellschaft	25,000 marks for an unlimited period of time
Kaiser-Wilhelm-Gesellschaft	25,000 marks for ten years
Koppel-Stiftung	25,000 marks for ten years

forschung ("for research on coal," 176,000 marks), and für experimentelle Therapie ("for experimental therapy," 94,000 marks) had larger budgets, whereas the KWI für Arbeitsphysiologie (for research on the physiology of work activities of the human body) received just over 40,000 marks per year. On the other hand, if compared with the university institutes, the KWI für Physik had a considerable amount of money at its disposal. Before the First World War, the average budget of a university's chemistry institute amounted to 37,000 marks and that of a physics institute to 14,500 marks per year.[31] In 1917, Planck's Institute for Theoretical Physics at the University of Berlin had one extraordinary professor (Max Born), one assistant, and a budget of only 700 marks. Nernst's Physicochemical Institute at the same university had, besides the director, four research staff, three other employees, and a 19,525 mark budget for non-personnel spending. The staff of the Physical Institute directed by Rubens and Arthur Wehnelt, the biggest university institute for physics in Germany, included besides the two directors, six assistants, one technician, and two other employees. The budget for non-personnel spending amounted to 30,274 marks per year.[32] Thus, Einstein's institute had more money to support research than all three physics institutes of the University of Berlin taken together (Table 2.2).

In a deviation from the structure laid out in 1914, the new institute was built on two bodies only: (a) the Kuratorium (administrative board) of six members, three of whom were to come from the KWG, two from the Koppel-Stiftung, and one from the Ministry of Education; and (b) the Direktorium (scientific steering board) the members of which would be Berlin physicists for the duration of the war. This structure, together with the by-laws taken from the original founding proposal, was provisional and intended only to launch the institute. The Senat of the KWG nominated its own members to the Kuratorium, the industrialist Wilhelm von Siemens, Nernst, and Planck as a substitute for the donor Franz Stock, who did not accept the position. The Ministry for Education chose as its representative Schmidt-Ott, who was at that time the Minister, and as his deputy, Krüss. The Koppel-Stiftung nominated Koppel and Haber. The Direktorium was to consist of Einstein and all five of the scientists who proposed the foundation of the institute: Haber, Nernst, Planck, Rubens and Warburg.[33] Consequently, there was a sizeable overlap between the two boards: Haber, Nernst and Planck sat on both. All members of the Direktorium were also members

---

[31]All the reported data are taken from vom Brocke (1990a, 87, 145–148). The sums given do not include the salaries of directors or professors.

[32]For the institutes of the University of Berlin, see Haushalt (1917, 141), Hoffmann (1984).

[33]Minutes of the Senat meeting, 6 July 1917, AMPG, I. Abt., Rep. 1 A, Nr. 1656, p. 1; Wendel (1975, 358).

Table 2.2 Yearly budgets of some research institutes around 1917

KWI für Arbeitsphysiologie	40,000 marks
KWI für Biologie	140,000 marks
KWI für Chemie	133,700 marks
KWI für experimentelle Therapie	94,000 marks
KWI für Kohlenforschung	176,000 marks
KWI für Physik	75,000 marks
KWI für Physikalische Chemie und Elektrochemie	100,000 marks
Institute for Physical Chemistry, University of Berlin	19,525 marks
Institute for Physics, University of Berlin	30,274 marks
Institute for Theoretical Physics, University of Berlin	700 marks
University Institute for Chemistry (average)	37,000 marks
University Institute for Physics (average)	14,500 marks

of the Preussische Akademie der Wissenschaften.[34] On 26 November 1917 the first joint constitutive meeting of the boards took place under the presidency of Harnack in his office at the State Library: per the founding proposal, the purviews of the two boards were such that "the scientific initiative and its consequences are entrusted" to the Direktorium while the Kuratorium "has to supervise the budget and therefore each appropriation must be submitted to it for approval until a yearly budgetary plan is presented—to the extent that such a plan is feasible at all."[35] Siemens and Einstein were elected Chairmen of the Kuratorium and Direktorium, respectively.

Table 2.3 Direktorium of the KWI für Physik until 1931

Albert Einstein, Director	From 1917 (de facto resigned in 1922)
Fritz Haber	From 1917
Max von Laue	From 1921, Deputy Director from 1922
Walther Nernst	From 1917
Friedrich Paschen	From 1924
Max Planck	From 1917
Heinrich Rubens	1917–1922
Emil Warburg	1917–1931

---

[34]Haber had been elected to the Academy in November 1914.

[35]Minutes of the meeting, AMPG, I. Abt., Rep. 1 A, Nr. 1656, p. 9; Kirsten and Treder (1979, 150, vol. 1).

In the following years, several changes took place in the composition of the two boards (Tables 2.3 and 2.4). Following a suggestion by Einstein, Laue became a member of the Direktorium in December 1921 and Deputy Director of the institute in October 1922.[36] Of the original Direktorium members, Rubens died in June 1922 and Warburg in July 1931. In October 1924, the Direktorium elected as a new member Friedrich Paschen, who had just become President of the Physikalisch-Technische Reichsanstalt.[37] As to the Kuratorium, after the death of Siemens in October 1919, the KWG appointed Schmidt-Ott, who was no longer Prussian Minister of Education. Schmidt-Ott became chairman of the Kuratorium after the interim chairmanship of Planck in March 1919.[38] In December 1921, the by-laws of the KWG were changed which led to increased participation of both the National and the Prussian governments in the leading boards of the KWG and its institutes. Thus, besides Krüss representing the Prussian Ministry of Education, the high ministerial official Max Donnevert also entered the Kuratorium as representative of the National Ministry of the Interior.[39] In addition, since 1919 the managing director of the KWG, Friedrich Glum, was also present at the meetings.[40] In April 1931, at the suggestion of the KWG, Planck took over the chairmanship of the Kuratorium from Schmidt-Ott. The KWG also nominated, as additional Kuratorium members, the physicists James Franck (Göttingen), Wilhelm Hausser (Heidelberg), Heinrich Konen (Bonn), and Jonathan Zenneck (Munich), so that the dominance of the Berlin scientists was reduced.[41] A month later, at the suggestion of Planck, the KWG nominated as further Kuratorium members Erwin Schrödinger (Berlin) and the industrialist Albert Vögler.[42]

On 16 December 1917 the establishment of the KWI für Physik was made public in the advertising section of the Berlin newspaper *Vossische Zeitung*:

On October 1, the Kaiser-Wilhelm-Institut für physikalische Forschung was born. Its task is to initiate and support the systematically planned pursuit of important and urgent physical problems by gaining and funding especially qualified researchers.

The selection of problems, methods, and laboratories is made by the undersigned members of the Direktorium. Nevertheless, suggestions made to the Direktorium by other physicists will also be examined and, if approved, the proposed projects will be supported.

---

[36] See page 86.

[37] Minutes of the Direktorium meeting, 15 October 1924, AMPG, I. Abt., Rep. 1 A, Nr. 1661.

[38] Minutes of the Senat meeting, 28 October 1919, AMPG, I. Abt., Rep. 1 A, Nr. 1657, p. 99a; Schmidt-Ott to Planck, 2 March 1920, AMPG, I. Abt., Rep. 1 A, Nr. 1657, p. 120.

[39] Vom Brocke (1990b, 211); von Harnack to Schmidt-Ott, 7 April 1922, GStA, I. HA, Rep. 92 Schmidt-Ott, C, Bd. 84, p. 493.

[40] On Glum and his activities in the KWG, see vom Brocke (1990b, 251–266), Glum (1964, 229–339).

[41] Circular letter of the KWG, 20 April 1931, AMPG, I. Abt., Rep. 1 A, Nr. 1655, pp. 3–5.

[42] Minutes of the Senat meeting, 12 May 1931, AMPG, I. Abt., Rep. 1 A, Nr. 71, p. 365.

Although the institute can of course become fully effective only after the end of the war, nonetheless work is to be started now. Inquiries for detailed information should be addressed to the co-signer and chairman of the Direktorium, Professor Einstein (Haberlandstr. 5, Berlin-Schöneberg).

The Direktorium.

Einstein. Haber. Nernst. Rubens. Warburg.[43]

Table 2.4   Kuratorium of the KWI für Physik until 1931

Max Donnevert (nominated by the National Ministry of the Interior)	From 1922
James Franck (nominated by the KWG)	From 1931
Friedrich Glum (nominated by the KWG)	From 1919
Fritz Haber (nominated by the Koppel-Stiftung)	From 1917
Wilhelm Hausser (nominated by the KWG)	From 1931
Heinrich Konen (nominated by the KWG)	From 1931
Leopold Koppel (nominated by the Koppel-Stiftung)	From 1917
Hugo A. Krüss (nominated by the Prussian Ministry of Education)	From 1917
Walther Nernst (nominated by the KWG)	From 1917
Max Planck (nominated by the KWG)	From 1917, Chairman from 1931
Friedrich Schmidt-Ott (nominated by the Prussian Ministry of Education, later by the KWG)	From 1917 Chairman 1919–1931
Erwin Schrödinger (nominated by the KWG)	From 1931
Wilhelm von Siemens (nominated by the KWG)	1917–1919, Chairman
Albert Vögler (nominated by the KWG)	From 1931
Jonathan Zenneck (nominated by the KWG)	From 1931

The address given is Einstein's private one, confirming the fact that the new institute had no building of its own. Einstein evidently preferred to work at home. The fact that the institute's launch was advertised in just this way is not trivial. The wording had even been discussed in a specially convened meeting of the Direktorium.[44] Clearly, from the outset, the founders had a two-sided strategy in mind. On the one side, the leadership the Direktorium would exert by taking the initiative and determining the research program was emphasized. On the other side, the institute declared

---

[43]*Vossische Zeitung*, 16 December 1917, Morgen-Ausgabe, Nr. 641, 2. Beilage "Finanz- und Handelsblatt," [p. 3]. The same announcement appeared in the *Berliner Tageblatt*, 20 December 1917, evening edition; and in the *Frankfurter Zeitung*, 21 December 1917, second morning edition. Planck's name is missing under the announcement because Einstein forgot it (see Planck to Einstein, 29 December 1917, in Einstein (1998, doc. 423)). The institute's opening was also later made public in various journals: *Elektrotechnische Zeitschrift*, 39 (1918), p. 179; *Die Naturwissenschaften*, 6 (1918), pp. 34–35; and *Physikalische Zeitschrift*, 19 (1918), p. 16.

[44]Planck to von Harnack, 16 November 1917, AMPG, I. Abt., Rep. 1 A, Nr. 1656, p. 6.

Fig. 2.1 Announcement of
the establishment of the KWI
für Physik, *Vossische
Zeitung*, 16 December 1917

Am 1. Oktober 1917 ist das

**Kaiser-Wilhelm-Institut
für physikal. Forschung**

ins Leben getreten. Seine Aufgabe soll
darin bestehen, die planmäßige Bearbei-
tung wichtiger und dringlicher phys!-
kalischer Probleme durch Gewinnung
und materielle Unterstützung besonders
geeigneter Forscher zu veranlassen und
zu fördern.

Die Auswahl der Probleme, der Me-
thoden sowie des Arbeitsplatzes liegt in
der Hand des unterzeichneten Direk-
toriums. Doch sollen auch von anderen
Physikern an das Direktorium gelangende
Anregungen von diesem erwogen und
die vorgeschlagenen Untersuchungen im
Falle der Billigung gefördert werden.

Wenn das Institut auch naturgemäß
erst nach Beendigung des Krieges seine
volle Wirksamkeit wird entfalten können,
so soll doch womöglich schon jetzt mit
der Arbeit begonnen werden. Angaben
über nähere Einzelheiten sind an den
mitunterzeichneten Vorsitzenden des Di-
rektoriums, Professor Einstein (Haberlandstr 5, Berlin-Schöneberg) zu richten.

Das Direktorium.
Einstein. Haber. Nernst. Rubens. Warburg.

its readiness to consider external suggestions. This was very unusual in that none of
the other KW Institutes had publicly appealed to the scientific community for help in
deciding the direction of their research. But precisely this open-mindedness towards
what was going on in the physics community and the awareness that cooperation was
needed had been the mainspring for the institute's foundation and specific mode of
organization.[45] In the announcement, no hint was given to the extent of the support
of either experimental or theoretical physics (Fig. 2.1).

---

[45]Two years later, concluding a speech at the general meeting of the KWG, Planck restated that
the KWI für Physik was doing pioneering work at the forefront of physics and was therefore open
to external suggestions. "The name of its director Albert Einstein guarantees that every suggestion
[...] promising some success would be most carefully examined and eventually strongly supported"
(Planck 1919, 909). In the version later published in a collection of speeches, Einstein's name is
omitted. Planck probably did not want to offend Nazi sensibilities (Planck 1943, 51).

References

Auer, Barbara (1984). *Das physikalische Institut in Heidelberg.* Heidelberg: Kunsthistorisches Institut der Universität Heidelberg.

Burchardt, Lothar (1975). *Wissenschaftspolitik im Wilhelminischen Deutschland. Vorgeschichte, Gründung und Aufbau der Kaiser-Wilhelm-Gesellschaft zur Förderung der Wissenschaften.* Göttingen: Vandenhoeck & Ruprecht.

Burchardt, Lothar (1990). "Die Kaiser-Wilhelm-Gesellschaft im Ersten Weltkrieg (1914–1918)". In: *Forschung im Spannungsfeld von Politik und Gesellschaft. Geschichte und Struktur der Kaiser-Wilhelm-/Max-Planck-Gesellschaft.* Ed. by Rudolf Vierhaus and Bernhard vom Brocke. Stuttgart: Deutsche Verlagsanstalt, pp. 163–196.

Einstein, Albert (1914a). "Beiträge zur Quantentheorie". In: *Verhandlungen der Deutschen Physikalischen Gesellschaft* 16, pp. 820–828. Reprinted in (Einstein 1996, doc. 5).

Einstein, Albert (1914b). "Die formale Grundlage der allgemeinenRelativitätstheorie". In: *Sitzungsberichte der Königlich Preussischen Akademie der Wissenschaften,* pp. 1030–1085. Reprinted in (Einstein 1996, doc. 9).

Einstein, Albert (1914c). "Kovarianzeigenschaften der Feldgleichungen der auf die verallgemeinerte Relativitätstheorie gegründeten Gravitationstheorie". In: *Zeitschrift für Mathematik und Physik* 63, pp. 215–225. Reprinted in (Einstein 1996, doc. 2).

Einstein, Albert (1915a). "Die Feldgleichungen der Gravitation". In: *Sitzungsberichte der Königlich Preussischen Akademie der Wissenschaften,* pp. 844–847. Reprinted in (Einstein 1996, doc. 25).

Einstein, Albert (1915b). "Erklärung der Perihelbewegung des Merkur aus der allgemeinen Relativitätstheorie". In: *Sitzungsberichte der Königlich Preussischen Akademie der Wissenschaften,* pp. 831–839. Reprinted in (Einstein 1996, doc. 24).

Einstein, Albert (1915c). "Zur allgemeinenRelativitätstheorie". In: *Sitzungsberichte der Königlich Preussischen Akademie der Wissenschaften,* pp. 778–786. Reprinted in (Einstein 1996, doc. 21).

Einstein, Albert (1915d). "Zur allgemeinen Relativitätstheorie (Nachtrag)". In: *Sitzungsberichte der Königlich Preussischen Akademie der Wissenschaften,* pp. 799–801. Reprinted in (Einstein 1996, doc. 22).

Einstein, Albert (1916). "Strahlungs-Emission und -Absorption nach der Quantentheorie". In: *Verhandlungen der Deutschen Physikalischen Gesellschaft* 18, pp. 318–323. Reprinted in (Einstein 1996, doc. 34).

Einstein, Albert (1917a). "Kosmologische Betrachtungen zur allgemeinen Relativitätstheorie". In: *Sitzungsberichte der Königlich Preussischen Akademie der Wissenschaften,* pp. 142–152. Reprinted in (Einstein 1996, doc. 43).

Einstein, Albert (1917b). "Zum Quantensatz von Sommerfeld und Epstein". In: *Verhandlungen der Deutschen Physikalischen Gesellschaft* 19, pp. 82–92.

Einstein, Albert (1996). *The Collected Papers.* Vol. 6: *The Berlin Years: Writings, 1914–1917.* Ed. by Anne J. Kox, Martin J. Klein, and Robert Schulmann. Princeton: Princeton University Press.

Einstein, Albert (1998). *The Collected Papers.* Vol. 8: *The Berlin Years: Correspondence, 1914–1918.* Princeton: Princeton University Press.

Einstein, Albert and Wander J. de Haas (1915). "Experimenteller Nachweis der Ampèreschen Molekularströme". In: *Verhandlungen der Deutschen Physikalischen Gesellschaft* 17, pp. 152–170. Reprinted in (Einstein 1996, doc. 13).

Fölsing, Albrecht (1993). *Albert Einstein. Eine Biographie.* Frankfurt am Main: Suhrkamp.

Forman, Paul (1968). *The Environment and Practice of Atomic Physics in Weimar Germany.* Ann Arbor (Michigan): UMI.

Glum, Friedrich (1964). *ZwischenWissenschaft,Wirtschaft und Politik. Erlebtes und Erdachtes in vier Reichen.* Bonn: Bouvier.

Goenner, Hubert and Giuseppe Castagnetti (1996). "Albert Einstein as Pacifist and Democrat during World War I". In: *Science in Context* 9, pp. 325–386.

Haushalt (1917). "Haushalt des Ministeriums der geistlichen und Unterrichts-Angelegenheiten für das Rechnungsjahr 1917". In: *Anlagen zum Staatshaushaltsplan für das Rechnungsjahr 1917* [Prussia]. Vol. 2. 31. Berlin: Reichsdruckerei.

Heisenberg, Werner (1971). "Das Kaiser-Wilhelm-Institut für Physik. Geschichte eines Instituts". In: *Jahrbuch der Max-Planck-Gesellschaft zur Förderung der Wissenschaften*, pp. 46–89.

Henning, Eckart (1996). "Das Harnack-Haus in Berlin-Dahlem". *Berichte und Mitteilungen 2/1996*. Munich: Max-Planck-Gesellschaft.

Hentschel, Klaus (1992a) *Der Einstein-Turm: Erwin F. Freundlich und die Relativitätstheorie*. Heidelberg: Spektrum Akademischer Verlag.

Hentschel, Klaus (1992b). "Einstein's Attitude towards Experiments: Testing Relativity Theory 1907–1927". In: *Studies in History and Philosophy of Science* 23, pp. 593–624.

Hentschel, Klaus (1994). "Erwin Finlay Freundlich and Testing Einstein's Theory of Relativity". In: *Archive for History of Exact Sciences* 47, pp. 143–201.

Hermann, Armin (1994). *Einstein. Der Weltweise und sein Jahrhundert*. Munich: Piper.

Hoffmann, Dieter (1980). "Albert Einstein und die Physikalisch-Technische Reichsanstalt". In: *Berliner Wissenschaftshistorische Kolloquien I: Über das persönliche und wissenschaftliche Wirken von Albert Einstein und Max von Laue*. Berlin: Akademie der Wissenschaften der DDR. Institut für Theorie, Geschichte und Organisation der Wissenschaft, pp. 90–102.

Hoffmann, Dieter (1984). "Die Physik an der Berliner Universität in der ersten Hälfte unseres Jahrhunderts. Zur personellen und institutionellen Entwicklung sowie wichtige Beziehungen mit anderen Institutionen physikalischer Forschung in Berlin". In: *Berliner Wissenschaftshistorische Kolloquien VIII: Die Entwicklung der Physik in Berlin*. Berlin: Akademie der Wissenschaften der DDR. Institut für Theorie, Geschichte und Organisation der Wissenschaft, pp. 5–29.

Janssen, Michael and Jürgen Renn (2007). "Untying the Knot: How Einstein Found His Way Back to Field Equations Discarded in the Zurich Notebook". In: *The Genesis of General Relativity*. Vol. 2. Ed. by Jürgen Renn. Dordrecht: Springer, pp. 839–925.

Jungnickel, Christa and Russell McCormmach (1986). *Intellectual Mastery of Nature. Theoretical Physics from Ohm to Einstein*. Chicago: The University of Chicago Press.

Kant, Horst (1987). "Das KWI für Physik—von der Gründung bis zum Institutsbau". In: *Berliner Wissenschaftshistorische Kolloquien XII: Beiträge zur Astronomie- und Physikgeschichte*. Berlin: Akademie der Wissenschaften der DDR. Institut für Theorie, Geschichte und Organisation der Wissenschaft, pp. 129–141.

Kant, Horst (1989). "Denkschriften für ein Kaiser-Wilhelm-Institut für Physik in Berlin". In: *Wissenschaft und Staat. Denkschriften und Stellungnahmen von Wissenschaftlern als Mittel wissenschaftspolitischer Artikulation*. Berlin: Akademie der Wissenschaften der DDR. Institut für Theorie, Geschichte und Organisation der Wissenschaft, pp. 165–183.

Kant, Horst (1993). "Peter Debye und das Kaiser-Wilhelm-Institut für Physik in Berlin". In: *Naturwissenschaften und Technik in der Geschichte. 25 Jahre Lehrstuhl für Geschichte der Naturwissenschaft und Technik am Historischen Institut der Universität Stuttgart*. Ed. by Helmuth Albrecht. Stuttgart: Verlag für Geschichte der Naturwissenschaften und der Technik, pp. 161–177.

Kant, Horst (1996). "Albert Einstein, Max von Laue, Peter Debye und das Kaiser-Wilhelm- Institut für Physik in Berlin (1917–1939)". In: *Die Kaiser-Wilhelm-/Max-Planck-Gesellschaft und ihre Institute*. Ed. by Bernhard vom Brocke and Hubert Laitko. Berlin: de Gruyter, pp. 227–243.

Kant, Horst (1992). "Institutsgründung in schwieriger Zeit. 75 Jahre Kaiser-Wilhelm-/Max-Planck-Institut für Physik". In: *Physikalische Blätter* 48, pp. 1031–1033.

Kerszberg, Pierre (1989). *The Invented Universe. The Einstein-De Sitter Controversy (1916–17) and the Rise of Relativistic Cosmology*. Oxford: Clarendon Press.

Kirsten, Christa and Hans-Jürgen Treder, eds. (1979). Albert Einstein in Berlin 1913–1933. Akademie-Verlag, Berlin

Kragh, Helge (1999). *Quantum Generations. A History of Physics in the Twentieth Century*. Princeton: Princeton University Press.

Mehra, Jagdisch and Helmut Rechenberg (1982). *The Historical Development of Quantum Theory*. Vol. 1: *The Quantum Theory of Planck, Einstein, Bohr and Sommerfeld: Its Foundation and the Rise of Its Difficulties 1900–1925*. New York: Springer.

Nathan, Otto and Heinz Norden (1968). *Einstein on Peace*. New York: Schoken.

Planck, Max (1919). "Das Wesen des Lichts". In: *Die Naturwissenschaften* 7, pp. 903–909.

Planck, Max (1943). *Wege zur physikalischen Erkenntnis. Reden und Vorträge*. Leipzig: Hirzel.

Renn, Jürgen and Tilman Sauer (2007) "Pathways Out of Classical Physics: Einstein's Double Strategy in His Search for the Gravitational Field Equation". In: *The Genesis of General Relativity*. Vol 1. Dordrecht: Springer, pp. 113–312.

Rubens, Heinrich and Gerhard Hettner (1916). "Das langwellige Wasserdampfspektrum und seine Deutung durch die Quantentheorie". In: *Sitzungsberichte der Königlich Preussischen Akademie der Wissenschaften*, pp. 167–183.

Schlüter, Steffen (1994). "Albert Einstein als Direktor des Kaiser-Wilhelm-Instituts für Physik". MA thesis. Berlin: Institut für Geschichtswissenschaft, Bereich Archivwissenschaft, Humboldt-Universität.

Schlüter, Steffen (1995). "Albert Einstein als Direktor des Kaiser-Wilhelm-Instituts in Berlin-Schöneberg". In: *Jahrbuch für Brandenburgische Landesgeschichte* 46, pp. 169–185.

vom Brocke, Bernhard (1990a). "Die Kaiser-Wilhelm-Gesellschaft im Kaiserreich. Vorgeschichte, Gründung und Entwicklung bis zum Ausbruch des Ersten Weltkriegs". In: *Forschung im Spannungsfeld von Politik und Gesellschaft. Geschichte und Struktur der Kaiser-Wilhelm-/Max-Planck-Gesellschaft*. Ed. by Rudolf Vierhaus and Bernhard vom Brocke. Stuttgart: Deutsche Verlagsanstalt, pp. 17–162.

vom Brocke, Bernhard (1990b). "Die Kaiser-Wilhelm-Gesellschaft in der Weimarer Republik. Ausbau zu einer gesamtdeutschen Forschungsorganisation (1918–1933)". In: *Forschung im Spannungsfeld von Politik und Gesellschaft. Geschichte und Struktur der Kaiser-Wilhelm-/Max-Planck-Gesellschaft*. Ed. by Rudolf Vierhaus and Bernhard vom Brocke. Stuttgart: Deutsche Verlagsanstalt, pp. 197–355.

Warburg, Emil (1916). "Über den Energieumsatz bei photochemischen Vorgängen in Gasen. VI. Photolyse des Brennwasserstoffs". In: *Sitzungsberichte der Königlich Preussischen Akademie der Wissenschaften*, pp. 314–329.

Wendel, Günter (1975). *Die Kaiser-Wilhelm-Gesellschaft 1911–1914. Zur Anatomie einer imperialistischen Forschungsgesellschaft*. Berlin: Akademie-Verlag.

Chapter 3
The KWI für Physik Under Einstein's Directorship 1917–1922

Abstract This chapter presents a detailed study of the activities of the Kaiser-Wilhelm-Institut für physikalische Forschung up to 1922. It considers why the large-scale scientific research projects envisaged from the outset were soon replaced with a larger number of separate research projects. Both the accepted and the rejected proposals are discussed, as well as the annual reports issued at the end of each fiscal year. Albert Einstein relinquished the directorship in 1922, officially in March 1923, and his reasons for stepping down are also addressed.

Keywords X-rays · Spectroscopy · Radiation phenomena · Planck's quantum radiation · *Zeitschrift für Physik*

3.1   A First Phase of Scarce Activity (1917–1919)

At first, the organizational effort put into the new institute remained minimal, with the secretarial work being done in Einstein's flat by Ilse Einstein, the eldest daughter of Elsa Einstein and her first husband, Max Löwenthal.[1] There is no indication whatsoever that the Direktorium met formally in Einstein's flat. Most probably, the meetings took place in the rooms of the Academy after its sessions.[2] Their number was kept to a minimum: only one in 1919, three meetings in both 1920 and 1921 and four in 1922. The frequency of meetings increased only after Laue took over the managing directorship. Then, if necessary, the Direktorium convened as much as three times a month. The decisions taken by the Direktorium were conveyed by

---

[1] Fölsing (1993, 461–462). In December 1917, Ilse Einstein was engaged for 50 marks monthly as personal secretary to Albert Einstein for three half-days per week, (Einstein 1998, 570, doc. 409). On a postcard of 12 May 1918 to Ilse, at that time out of Berlin, Albert writes "Some things have in fact arrived for the institute, although nothing worth mentioning. I'll save it all up until you come, so that you feel right in your element. Then we'll piece together the awful balance viribus unitis." He signed "Dein Prinzipal [Your Principal]" (Einstein 1998, 758, doc. 536).

[2] See, e.g., minutes of the Direktorium meeting, 26 January 1922, AMPG, I. Abt., Rep. 34, Nr. 12. Almost all the Direktorium meetings took place on the same days as the Academy sessions (see calendar of the Academy sessions in Kirsten and Treder (1979, 232–251, vol. 2).

© The Author(s), under exclusive license to Springer Nature Switzerland AG 2020
H. Goenner and G. Castagnetti, *Establishing Quantum Physics in Berlin*,
SpringerBriefs in History of Science and Technology,
https://doi.org/10.1007/978-3-030-63122-2_3

Einstein to the Kuratorium in the form of notes containing the name of the recipient, a definition of his research project, and the sum given. No documents have been found reporting scientific discussions among the Direktorium members concerning the projects or the reasons for their approval.

The director had very little autonomy in spending the funds. Einstein paid for the advertisement himself, but at about the same time wrote to Siemens:

> I think that, toward achieving the simplest possible accounting, it would be beneficial if small sums were put regularly at my disposal for the defrayal of running disbursements. It would probably be best for me to have an institute cashbox at my home containing a few hundred marks and to render an account of it at specified intervals.[3]

With this little money and Ilse's help, Einstein started the institute.

Erwin Freundlich Becomes a Researcher at the KWI für Physik

At first the institute took only one initiative, not in the field of radiation or quantum physics, but in general relativity.[4] In February 1918 Freundlich was appointed for three years with the task of conducting "experimental and theoretical astronomic research for testing the theory of general relativity and related questions."[5] After years of struggle, the establishment of the new institute at last allowed Einstein to concentrate forces on the experimental testing of his gravitational theory. Freundlich left the Sternwarte and continued his research as a staff member of the KWI für Physik on the premises of the Astrophysikalisches Observatorium (Astrophysical Observatory) in Potsdam near Berlin (Hentschel 1992b, 45, 51–58). Here, after training in photographic techniques (Müller 1919, 250), he worked on methods to detect the redshift in the spectral lines of fixed stars and made some first observations[6] that greatly pleased Einstein.[7] He was also very busy developing new apparatuses of his own conception.[8]

First Applications for Grants

Reaction to the opening announcement was immediate but not of the kind desired. Aside from requests for further information and mail from obvious cranks, most of the incoming letters sought support for the development of technical inventions or for research in fields outside physics. The suggestions ranged from developing techniques for melting materials with a high melting point, to designing special

---

[3] Einstein to von Siemens, undated [December 1917], in Einstein (1998, 570, doc. 409). Einstein even needed the Kuratorium's permission to buy a typewriter for 900 marks (Einstein to Bank Mendelssohn, 22 April 1919, AMPG, I. Abt., Rep. 1 A, Nr. 1656, p. 52; and von Siemens to Bank Mendelssohn, 25 April 1919, AMPG, I. Abt., Rep. 34, Nr. 8, folder Mendelssohn).

[4] For a list of the research projects supported from the institute's inception until the budget year 1922/23, see Appendix.

[5] Contract between KWI für Physik and Freundlich, 4 February 1918, AMPG, I. Abt., Rep. 34, Nr. 2, folder Freundlich.

[6] Freundlich to Einstein, 27 March 1919 (Einstein 2004, 25–26, doc. 14).

[7] Einstein to Freundlich, 29 March 1919 (Einstein 2004, 27, doc. 15).

[8] Freundlich to Einstein, 31 October 1918, AMPG, I. Abt., Rep. 34, Nr. 2, folder Freundlich.

machines for use in coal mines, to the development of procedures to manufacture hollow structures made of concrete.[9]

Of particular interest were the letters from Hermann Fricke, an examiner at the Patent Office who in the following years wrote many books and articles against Einstein's theory of general relativity. In his first letter Fricke asked Einstein to referee for publication an enclosed paper expounding his pretended "discovery" of a proportional relation between the temperature of the solar surface and gravitation. Fricke hoped for an unbiased examination of his theory. Einstein answered probably with an elaborate rejection of which, unfortunately, only one sentence quoted by Fricke in his last letter remains: "According to all we know, the release of energy by celestial bodies in proportion to the product of surface and absolute temperature seems totally impossible."[10]

Only two proposals were worthy of consideration. Dr. med. Gustav Bucky, of the X-ray department at the Practical School for Public Health Service of the Berlin University, wanted to do research on the reaction of biological systems to X-rays and suggested the establishment of "a special department for radiological research." As he wrote:

> [W]e medical doctors are convinced that mainly the theoretical foundations and, in particular, the precise measurement of radiation according to quantity and quality are absolutely necessary for fruitful work. Without this foundation, it is impossible to gain a clear insight into the healing effect, and [...] to advance the methodology of irradiations.[11]

Apparently, Einstein was not in favor of adding a department for radiation research to his institute, but Bucky insisted and asked Einstein's opinion about an "intermediary institute" in which topics "at the borderline between physics and physiology" would be systematically investigated "by the joint work of physicists, zoologists, botanists, and physicians." Topics of "electromedicine" could also be the subject of research. In Bucky's opinion, industry would be very interested in such investigations.[12] We do not know Einstein's response, but Bucky's plan did not gain the Direktorium's approval because it clearly did not fall into the category of research it wanted to support, that is, fundamental theoretical and experimental research on the molecular theory of matter—despite the vague reference to cancer research recurring in the documents prior to the institute's foundation.

Only the project submitted by Peter Debye corresponded particularly well with the declared intentions of the KWI für Physik. Debye was at the time director of the Physical Institute of the University of Göttingen. Einstein held Debye in high esteem and had even recommended him as his successor at the University of Zurich in 1911 (Einstein 1993b, 290–291). Debye now asked Einstein for a grant of 16,030 marks

---

[9]Einstein (1998, 1013–1030). Einstein's answers are missing.

[10]Quotation from Fricke to Einstein, 10 February 1919, AMPG, I. Abt., Rep. 34, Nr. 2, folder Fricke. See also Fricke to Einstein, 18 February and 15 March 1918, ibid.; (Einstein 1998, 1018–1019; Goenner 1993, 127–128). For Fricke's bibliography, see Hentschel (1990, xxx–xxxi).

[11]Bucky to Einstein, 11 May 1918, AMPG, I. Abt., Rep. 34, Nr. 1, folder Bucky.

[12]Bucky to Einstein, 18 May 1918, ibid. Einstein and Bucky became friends for life and later developed a joint patent on a shutter for X-ray cameras (Bucky 1991, 12–15).

to buy instruments for generating "X-rays of arbitrary wave length and of sufficient intensity." He had spoken about his research with Born and Sommerfeld and had been advised by the Prussian Minister of Education at that time, Schmidt-Ott, to apply to the KWI für Physik and to Einstein.[13] In his application Debye enclosed a paper concerning the structure of the atom in crystals, in particular the number and position of electrons and the "size of the planetary system of electrons belonging to the atom" (Debye and Scherrer 1918, 120). While at the beginning of the paper the question was posed whether chemical valency is related to Planck's quantum of action, the authors noted at the end that:

> [A]ll the considerations were made with the help of classical foundations. We do not know of any experience which would put into doubt the conclusions drawn from these foundations in the range of wavelengths we have used. However, for much smaller wavelengths in the range of γ-rays observations of absorption seem to point to new aspects. (Debye and Scherrer 1918, 120)

Debye took up these new aspects in his proposal. The method of investigation he suggested made use of X-ray scattering off a crystal (diamond). The incoming ray would be absorbed by the electrons and then reradiated. From intensity measurements "inferences can be made about the number of electrons and the distance between them" in the atom. As hinted at in the paper, Debye was particularly interested in X-rays with very short wavelengths, that is, high energies, for which he expected "a fault in classical electrodynamics. (I am thinking here of a quantization of the radiation emitted from a free electron)."[14]

Einstein received Debye's application during his vacation at the Baltic Sea. Probably assuming that the other Direktorium members were still in Berlin, he forwarded the letter to Planck adding a short note in which he expressed strong support, without commenting on the scientific aspects:

> The letter speaks for itself. I believe that there could be no better use of our money than by placing the desired apparatus at Debye's disposal (purchase and loan them out to him for as long as he would like to have them. [...] we have just *one* Debye and his life span is $< \infty$.[15]

He then concluded: "I ask you now to convene a meeting of the Direktorium as soon as possible and to confer on the matter. Then I request a brief report on the outcome, so that I can negotiate further with Debye." Planck, after expressing his satisfaction with the fact that "the KWI für Physik gets such a fine opportunity to demonstrate its usefulness," gave a mild but polite rebuke: "But I believe that for the moment we can do without a Direktorium meeting (which, incidentally, I am not at all authorized to convene); because the Direktorium members will surely be grateful for not becoming involved in person." Presumably, they were also on vacation. Planck then proposed handling the matter in the same way as with Freundlich. He suggested

---

[13]Debye to Einstein, 2 July 1918, in Einstein (1998, doc. 577).

[14]Debye to Einstein, 2 July 1918, in Einstein (1998, 820–822, doc. 577). On Debye's research project, see also Kant (1993, 163–164).

[15]Einstein to Planck, undated, addition to the letter of Debye to Einstein, 2 July 1918, in Einstein (1998, 823, doc. 578). Emphasized text is underlined in the original by Einstein.

that Einstein draft a contract with Debye which the other members would approve by mail.[16] And so it happened without disagreement.[17]

Planck, bearing in mind that the institute intended to initiate research of its own, remarked: "Pondering the fact that we are concerned here with a good and promising project with every guarantee for a very valuable scientific success, we must put aside the reservation that the KWI appears here only as a source of money."[18] Nernst also gave his signature with an interesting comment: "I [...] agree all the more, as it was originally a program in the same direction as the one now proposed by Debye that led to the establishment of the [Kaiser Wilhelm] Institute for Physics."[19] While Planck and Nernst hinted at the need for a rationale for the institute, Director Einstein seemed satisfied with shuffling the papers.

Debye received the money he had applied for roughly three months later. As it turned out, the manufacturer Siemens could not provide the needed high-voltage transformer until summer 1920, and then for a price which had quadrupled due to post-war inflation. When the instrument was sent to Göttingen, Debye had already left for an appointment in Zurich. In the end, he returned the grant plus the accumulated interest.[20]

The First Annual Report

During 1918 the institute did not financially support any other research projects, though not because of a lack of money. As already reported, the institute had an annual budget of 75,000 marks.[21] Einstein received an annual remuneration of 5,000 marks, his secretary Ilse 600 marks, and Freundlich 6,000 marks. At the end of 1918, the bank statement showed a balance of 66,156 marks.[22] Three months later, at the end of the budgetary year 1918/19, the institute still had a surplus of 82,374 marks. Only 28,435 marks had been spent, including salaries, running costs, the grant for

---

[16]Planck to Einstein, 8 July 1918, in Einstein (1998, 830, doc. 584).

[17]Contract between KWI für Physik and Debye, 27 August 1918, AMPG, I. Abt., Rep. 34, Nr. 1, folder Debye; Einstein to Planck, 19 July 1918, AMPG, I. Abt., Rep. 1 A, Nr. 1656, p. 35. See also Einstein (1998, 1025).

[18]Addition by Planck, 21 July 1918, to the letter of Einstein to Planck, 19 July 1918, AMPG, I. Abt., Rep. 1 A, Nr. 1656, p. 35.

[19]Addition by Nernst, 24 July 1918, see footnote 18.

[20]Debye to Einstein, 7 June 1920, 20 December 1920, and 1 February 1921, all in AMPG, I. Abt., Rep. 34, Nr. 1, folder Debye; von Siemens to Bank Mendelssohn, 10 October 1918, AMPG, I. Abt., Rep. 34, Nr. 8, folder Mendelssohn.

[21]See Table 2.1. For a survey of incomes and expenditures until the budget year 1922/23, see Table 3.1. Table 3.2 gives the repartition of expenditures. The sums of the tables are in marks and calculated on the basis of the available documents. They should be considered only as indicative; due to discrepancies and errors in the administrative documents (see, e.g., the survey for 1921/22, AMPG, I. Abt., Rep. 1 A, Nr. 1665, p. 59), it is not possible to reconstruct the actual situation with certainty.

[22]Bank statement 1 April–31 December 1918, AMPG, I. Abt., Rep. 34, Nr. 13.

Table 3.1 Survey of incomes and expenditures

Budgetary year	Previous balance	Income	Expenditure	New balance
1917/18	0	37,891	4,693	33,198
1918/19	33,198	77,611	28,435	82,374
1919/20	82,374	78,111	76,559	83,926
1920/21	83,926	146,845	89,476	141,295
1921/22	141,295	164,730	108,074	197,951
1922/23	197,951	923,572	918,521	203,002

Table 3.2 Repartition of expenditures

Budgetary year	Einstein's salary	Expenditure for scientific purposes (incl. Freundlich's salary)	Miscellaneous (incl. admin. personnel costs)	Total
1917/18	2,500	1,500	693	4,693
1918/19	5,000	22,530	905	28,435
1919/20	5,000	66,050	5,509	76,559
1920/21	10,000	76,100	3,376	89,476
1921/22	15,000	88,675	4,399	108,074
1922/23	33,505	879,013 (366,000 for the Deutsches Entomologisches Museum)	6,003	918,521

Debye, and 500 marks for reimbursement of expenses for Freundlich.[23] However, there were reasons for this restraint. First and foremost, almost all the people who could have been taken into consideration as grant recipients were either drafted into the armed forces or required to work towards the war effort in some other way. The war and the following revolution of November 1918, with the accompanying general turmoil, certainly did not allow the elaboration and implementation of a systematic research program. Finally, a long illness prevented Einstein from working for almost a year (Fölsing 1993, 462–463).

In the first annual report for the budgetary year 1918/1919 the situation becomes obvious:

---

[23] Survey of income and expenditure 1 April 1918–31 March 1919, AMPG, I. Abt., Rep. 1 A, Nr. 1665, pp. 18–19; bank statement 1 January–31 March 1919, ibid., Rep. 34, Nr. 8, folder Mendelssohn.

Since the number of colleagues who were able to devote themselves to goals in pure science during the past year was minimal because of the war, the largest amount of the money available for the budgetary year 1918/19 was set aside for the present one [1919/20]. Thus only two researchers were working for the institute.[24]

Einstein then reports about Freundlich's work concerning the displacement of both stellar and solar spectral lines and of Debye's investigation into the constitution of matter by X-rays.

3.2 The Real Start. The Period 1919/20

While the Direktorium had not officially convened since the constitutive meeting of November 1917, on 15 March 1919 it suddenly sent a letter to the heads of the physics institutes of all universities "of Germany and of German-speaking Austria" notifying them of the availability of means for financing research projects.

On the occasion of the new start of scientific research in physics, we would like to direct the attention of our colleagues to the fact that the Kaiser Wilhelm Institute for Physics has at its disposal considerable means, which can be given to scientific institutions as well as to individual colleagues in order to make possible or to facilitate scientific research. In particular, we shall consider

1. the purchase of instruments to be used for special scientific investigations;
2. fellowships for the pursuit of definite scientific research projects.

Proposals with statements explaining the planned scientific investigations as well as the funds needed are to be sent to Prof. Einstein [...]. The only criterion for the distribution of funds will be that of reviving physics research.[25]

In sharp contrast to the intentions declared in the press announcement of December 1917, the implementation of a planned specific research program of its own was no longer at the top of the institute's agenda. The money accumulated by this time was now to be distributed according to the suggestions coming from the physics community, with only the quality of the research to be refereed by the Direktorium, and not its subjects. The KWI für Physik was thus taking on a more general task of backing the state in supporting ongoing research. Actually, as we will see, the Direktorium decided to reserve a considerable amount of money to support research on a fundamental problem of quantum physics and to that extent did set its own

---

[24]Report of the KWI für Physik 1 April 1918–31 March 1919, AMPG, I. Abt., Rep. 1 A, Nr. 1665, p. 33. The report was sent to the Kuratorium in June 1919 (Einstein to Kuratorium, 16 June 1919, ibid., Nr. 1656, p. 64), after Siemens's complaints (see page 58).

[25]Direktorium of the KWI für Physik to "Kollege," AMPG, I. Abt., Rep. 34, Nr. 13. Some versions of this circular were probably dated "18 March 1919" (see, e.g., Lehmann to Einstein, 26 March 1919, AMPG, I. Abt., Rep. 34, Nr. 8, folder Lehmann; Steubing to Einstein, 6 April 1919, ibid., Nr. 10, folder Steubing) or "19 March" (see Jensen to Einstein, 14 May 1919, ibid., Nr. 5, folder Jensen).

priorities. Nevertheless, a shift from an initiating role to a more passive one is evident if we recall the repeated statements about the institute's aims in the founding documents. The reasons for this change are not easy to discern. The impression is that the Direktorium members were uncertain and looking around for ideas. In 1919, Einstein justified the new policy of broadly distributing funds by referring to the "difficulties with which the physics institutes [of the universities] have to struggle because of the economic situation,"[26] but, as we will argue later, this was probably not the entire truth.

Scientists Funded and Their Projects

In reaction to the Direktorium's invitation, in the following weeks the institute received twenty grant applications.[27] A first series of allocations was decided upon at a Direktorium meeting on 24 April 1919. The next day, Einstein sent the list of grants to the Kuratorium for approval, adding to each name and sum a very vague justification but no detailed explanation of the project.[28]

The largest sum, 20,000 marks, corresponding to a third of the total grants, was set aside "for a new precision measurement of the spectral law," that is, Planck's radiation law. Clearly, this meant that the question was considered of primary importance. In fact, the very day after the Direktorium meeting Nernst presented at a meeting of the Deutsche Physikalische Gesellschaft (German Physical Society) the results of his and Theodor Wulf's critical examination of the measurements taken since 1900 in support of Planck's law. Through new calculations they had obtained values for emissivity showing a small but systematic deviation from the values predicted by the law, thus coming to the conclusion that Planck's radiation formula was not a "strict natural law" and that a corrective factor should be added.[29] Of course, the Direktorium was well aware of the serious consequences that a discrepancy, however small, would have for the radiation theory and for quantum physics, which is based on the "absolute and exact validity of Planck's formula" (Hettner 1922, 1037). It was necessary to ascertain whether the discrepancy was due to measurement errors or instead revealed a real inadequacy of the formula. A new series of precise experimental observations had to be undertaken.

Two months after the Direktorium meeting, Einstein arranged that, from the money set aside for testing Planck's formula, 5,000 marks were to be remitted to Rubens and 4,000 marks to Warburg.[30] The delay between the decision to support research concerning Planck's law and the actual transfer of funds could mean that the Direktorium had decided immediately that the question was worth tackling, but

---

[26] Einstein to Kuratorium, 25 April 1919, AMPG, I. Abt., Rep. 1 A, Nr. 1656, pp. 55–56.

[27] It must be noted that the archives do not contain applications for every project supported, nor is there a response to every existing application.

[28] E.g., "for instruments for spectroscopic research" or "for spectroscopic research on X-rays," etc. (Einstein to Kuratorium, 25 April 1919, AMPG, I. Abt., Rep. 1 A, Nr. 1656, pp. 55–56). Examples of Direktorium dispositions are given by Kant (1987, 133). The motivations given in the Appendix are taken from the available correspondence.

[29] Nernst and Wulf (1919), see also Hettner (1922, 1037).

[30] Einstein to KWG, 27 June 1919, AMPG, I. Abt., Rep. 1 A, Nr. 1656, p. 77.

did not yet know who should perform the experiments. Perhaps Rubens and Warburg were first asked to look for suitable collaborators or to define what kind of instrumentation they needed for their respective institutes before a precise sum could be allocated. Both had already worked on the subject and their results had been examined by Nernst and Wulf in their paper. As is well known, the observations by Rubens and Ferdinand Kurlbaum into the variations of radiation intensity with temperature in the infrared spectrum had provided Planck experimental bases for the enunciation of his spectral law in 1900. Immediately thereafter, Rubens and Kurlbaum had contributed to the confirmation of Planck's law by performing measurements covering an extended range of wavelengths (Hettner 1922, 1035–1037; Kangro 1970, 200–207). As for Warburg, in the years 1913–1915 he had carried out with his coworkers a series of precision measurements of what they called "the constant c of the Wien-Planck radiation law," which is known today as the second radiation constant c_2 (Warburg et al. 1913; Warburg and Müller 1915).

Whereas Warburg soon reported that he could not do the intended research because of unspecified "hindrances,"[31] Rubens started, with his collaborator Gerhard Michel, a new series of extremely precise measurements of isochromates which, thanks to the new instruments he had developed, led to the confirmation of Planck's formula.[32] Rubens's work continued to be funded by the KWI für Physik in the following years and was mentioned by Einstein in the activity report for the year 1921/22 as one of the most important research projects supported by the institute, precisely because of its consequences for the foundations of quantum physics.[33]

Nernst was most probably the sponsor of a grant of 2,000 marks for Paul Günther, an assistant at the Institute for Physical Chemistry at Berlin University, "for thermal research on amorphous solids."[34] Concurrent with his examination of Planck's formula, Nernst had renewed his interest in his hypothesis of gas degeneracy at extremely low temperatures. In a recent paper he had derived from this hypothesis some consequences that could be detected experimentally (Nernst 1919). And in fact Günther spent the money—in accordance with Nernst's wishes, as he reported—for a viscosimeter and the pertaining liquid hydrogen used to perform measurements on the temperature dependence of the coefficient of viscosity that confirmed Nernst's prediction.[35]

---

[31] Warburg to Kuratorium of the KWI für Physik, 10 July 1919, AMPG, I. Abt., Rep. 1 A, Nr. 1656, p. 79.

[32] Rubens and Michel (1921a, b). The deviation calculated by Wulf and Nernst was therefore probably due to unprecise measurements, see also Hettner (1922, 1037–1038).

[33] Report of the KWI für Physik 1 April 1921–1931 March 1922, AMPG, I. Abt., Rep. 1 A, Nr. 1665, pp. 57–58.

[34] Einstein to Kuratorium, 25 April 1919, AMPG, I. Abt., Rep. 1 A, Nr. 1656, pp. 55–56; see also Einstein to Bank Mendelssohn, 22 May [recte June] 1919, ibid., pp. 67–68.

[35] Günther to KWI für Physik, 27 March 1924, AMPG, I. Abt., Rep. 34, Nr. 3, folder Günther; (Günther 1920, 1924b). See also the note concerning (Günther 1920) in *Sitzungsberichte der Preussischen Akademie der Wissenschaften*, 1920, p. 575.

Only the application by Leonhard Grebe and Albert Bachem, both *Privatdozenten* at the University of Bonn, had been solicited by Einstein.[36] He and Grebe had met in Berlin probably at the beginning of April 1919, and discussed a program of investigations designed to verify the gravitational redshift of the sun's spectral lines, which was considered to be evidence for general relativity. Grebe, a specialist in spectroscopy, had already been working for some months with Bachem on the cyanogen bands of the solar spectrum and found that while some lines showed "with sufficient approximation" the displacement predicted by the theory, others presented anomalies that could not be due to observational errors. In order to explain these anomalies Grebe and Bachem wanted to analyze the structure of the lines with a Koch microphotometer.[37] They received an initial sum of 2,000 marks from the KWI für Physik but evidently needed more, since Einstein later helped arrange a further 2,000 marks from a private donor, the publisher and philanthropist Richard Fleischer.[38] Still, the money was not enough to buy either the Koch instrument or a cheaper one, so it was eventually spent on research visits, in summer 1919 and during the Easter vacation 1920, to the Astrophysikalisches Observatorium in Postdam, where Grebe and Bachem used the Koch microphotometer installed by Freundlich.[39]

Judging from the existing correspondence and in comparison with the slight interest he showed for other research supported by his institute, Einstein followed Grebe and Bachem's work with exceptional attention. As early as June 1919, although the required analysis had not yet been done, Einstein pressured them to publish a paper on the preliminary research, which could at least support the conclusion that "the existence of the Einstein effect receives a high degree of probability."[40] After an initial investigation performed in Potsdam, Grebe and Bachem sent Einstein a paper in which they claimed that "the Einstein gravitational shift in the solar field is really there both as to the direction and the amount" (Grebe and Bachem 1920b, 54). This conclusion was essentially based on the data of a particular, small group of spectral lines and a great part of the paper was dedicated to justifying the selection criteria.

---

[36]We cannot reconstruct how Einstein became acquainted with Grebe who, with his application, was reacting to an "Aufforderung" (invitation) by Einstein.

[37]Grebe to Einstein, 17 April 1919, (Einstein 2004, 37–38, doc. 25). The research results are presented in Grebe and Bachem (1919, 1920a, b), and in Grebe (1920, 1921). A detailed account of Grebe and Bachem's work is given by Hentschel (1992c); see also Earman and Glymour (1980, 194–196), Hentschel (1998, 514–535).

[38]Einstein to Kuratorium, 25 April 1919, AMPG, I. Abt., Rep. 1 A, Nr. 1656, pp. 55–56; Fleischer to Einstein, 21 and 29 December 1919 (Einstein 2004, 319, doc. 227 and 331–332 doc. 238, respectively); Einstein to Fleischer, 17 January 1920, quoted in L'Autographe (1992, 18, lot 42). On Fleischer, see Starkulla (1971).

[39]Grebe to Einstein, 6 June 1919 (Einstein 2004, 86–87, doc. 57); Grebe and Bachem to Einstein, 26 January 1920 (Einstein 2004, 385–386, doc. 283); Einstein to Grebe, 9 July 1920, AMPG, I. Abt., Rep. 34, Nr. 3, folder Grebe; Grebe to Einstein, 12 July 1920, AEA, 6–048; Grebe to Laue, 10 March 1924, AMPG, I. Abt., Rep. 34, Nr. 3, folder Grebe; Freundlich, "Bericht über die Tätigkeit im Jahre 1919," AMPG, I. Abt., Rep. 34, Nr. 2, folder Freundlich (Grebe and Bachem 1920b, 51).

[40]Grebe and Bachem (1919, 464); see also Grebe to Einstein, 29 June 1919, AMPG, I. Abt., Rep. 34, Nr. 3, folder Grebe.

Furthermore, since the first investigation had been disturbed by a technical defect in the Koch microphotometer, Grebe and Bachem planned to repeat the measurements and deferred the publication of the microphotometric pictures to a conclusive paper (Einstein 2004, 324–325 and 385–386). Nevertheless, Einstein immediately announced the favorable outcome to the physics community.[41] He also insisted on the publication of the observational results: "Without such publication the colleagues would hardly attribute any evidential value to your work."[42] Unfortunately, the second series of microphotometric measurements was also unsatisfactory because of other disturbing effects caused by the instrument. This led Grebe and Bachem to publish all the photograms together with the displacement data of the previous paper, while sticking to their conclusion (Grebe and Bachem 1920a). Although Einstein agreed on the narrow lines' selection, he was now less convinced of the conclusiveness of Grebe and Bachem's work which, in his words, "does not yet prove the existence of the redshift but makes it at least probable."[43] Nonetheless, Grebe presented his work on the gravitational shift in support of the theory of general relativity at the annual meeting of the Gesellschaft Deutscher Naturforscher und Aerzte in September 1920, where his talk was followed by a heated discussion between Einstein, Lenard, and others.[44]

In addition to Grebe and Bachem, in this fiscal year the KWI für Physik supported two other researchers working for Einstein. Freundlich received an additional 2,300 marks as a cost-of-living allowance and 550 marks for reimbursement of expenses, in particular for the installation of the Koch microphotometer which was accomplished in summer 1919.[45] During this year, Freundlich continued his systematic investigation of the displacement of solar and stellar spectral lines, which did not yet lead to conclusive results because of several technical and observational problems.[46] Most probably, Einstein himself requested 1,200 marks for Jakob Grommer as compensation for "mathematical work."[47] Grommer, a mathematician and former student of Hilbert without an academic position, is known to have been working with Einstein at least since 1916 because his help was acknowledged in the "Kosmologische Betrachtungen" of early 1917.[48] After that he collaborated with Einstein

---

[41] Hentschel (1992c, 33); see also, e.g., Einstein (2004, 342, 353).

[42] Einstein to Grebe, 9 July 1920, AMPG, I. Abt., Rep. 34, Nr. 3, folder Grebe.

[43] "Durch diese Arbeit wird die Existenz der Rotverschiebung zwar noch nicht sicher bewiesen, aber doch schon wahrscheinlich gemacht" (Einstein to Lorentz, 4 August 1920 (Einstein 2006, 364–365, doc. 98)). See also Einstein (2006, 337, 346).

[44] Grebe (1920). Grebe had been invited by Sommerfeld (Einstein and Sommerfeld 1968, 69).

[45] Einstein (2004, 107, 447, 591); Einstein to Kuratorium, 20 December 1919, AMPG, I. Abt., Rep. 1 A, Nr. 1657, 109.

[46] Freundlich, "Bericht über die Tätigkeit im Jahre 1919," undated [January 1920?], AMPG, I. Abt., Rep. 34, Nr. 2, folder Freundlich; see also Müller (1920, 117).

[47] Einstein to Kuratorium, 25 April 1919, AMPG, I. Abt., Rep. 1 A, Nr. 1656, pp. 55–56; (Einstein 2004, 560).

[48] Einstein (1917a, 146). In summer 1917, Einstein had even "authorized" Grommer to complete his lectures on relativity theory at the Berlin University during his absence (Einstein to Ehrenfest, 22 July 1917, in Einstein (1998, 484, doc. 362)). On Grommer, see Pais (1982, 487–488), Einstein

for more than ten years on problems relating to general relativity, quantum theory, and unified field theory. Between spring and early summer 1919 Grommer worked, certainly at Einstein's request, on a problem pertaining to the mathematical proof of the energy conservation law in the theory of general relativity. In his paper on the energy conservation law published in the previous year, Einstein had left one particular assumption in his derivation of the law without proof (Einstein 1918, 457). This proof was provided by Grommer in July 1919 in a letter to Einstein, in which he also acknowledges receipt of the institute's money.[49] Grommer's paper was then submitted by Einstein to the Prussian Academy for publication in November 1919 (Grommer 1919).

As to the other applications, most of them came either from well-known colleagues or from their assistants, but some were also from young scholars at the beginning of their careers. The fields of research were quite diverse. A considerable group of proposals was concerned with spectroscopy of single atoms or molecules. Both optical and X-ray line spectra reflect the structure of atoms, band spectra the internal degrees of freedom of molecules. In particular, with his refinement of Bohr's atomic model, Sommerfeld had been able to show how the series of line spectra of hydrogen and helium could be understood. The research proposals addressed new ideas as well as old questions.

First it is worth noting the great support given to Ernst Wagner, extraordinary professor of physics at the University of Munich. Wagner, a specialist on X-ray spectroscopy, had recently been able to obtain a more precise determination of Planck's constant h through an improvement of the experimental setup.[50] Now he needed a high-voltage battery in order to extend his systematic investigation, for which he was granted 10,000 marks.[51] The sum was one of the largest distributed that year. This fact, together with the support for Rubens's measurements concerning the spectral law, confirmed the interest of the Direktorium members in research in this field. Wagner's work continued to be supported in the following years.

Hugo Seemann, a guest scholar at the University of Würzburg (where he enjoyed the patronage of Wien), proposed measurements of the polarization of X-rays which, in principle, could test Sommerfeld's theory of "Bremsstrahlung," that is, continuous radiation emitted by decelerated electrons. Being deaf, he had neither a position nor a chance for an academic career, so he asked for a fellowship.[52] Nobel Prize laureate Laue also backed Seemann's application by pointing out his experimental skills and industriousness.[53] However, in violation of the conditions stated in the

(1998, 485, note 1). An autobiographical note by Grommer, probably written in November 1926, is preserved in the Bundesarchiv Koblenz, R 73, Nr. 16393.

[49] Grommer to Einstein (2004, 100–101, doc. 67).

[50] Wagner (1918). On Wagner, see Valentiner (1929), Mehra and Rechenberg (1982, 322, vol. 1, pt. 1).

[51] Wagner to Einstein, 5 April 1919, AMPG, I. Abt., Rep. 34, Nr. 11, folder Wagner; Einstein to Kuratorium, 25 April 1919, AMPG, I. Abt., Rep. 1 A, Nr. 1656, pp. 55–56; Einstein (2004, 560).

[52] Seemann to Einstein, 26 March 1919 (Einstein 2004, 21–24, doc. 13).

[53] Laue to Einstein, 7 April 1919 (Einstein 2004, 30–31, doc. 18).

March circular, the Direktorium decided not to give fellowships and instead granted 3,000 marks to buy unspecified "instruments for spectroscopic research on X-rays."[54] Privately, Einstein offered his help to obtain a position at the Physikalisch-Technische Reichsanstalt in Berlin, but Seemann refused because at the Reichsanstalt he would not have been free to pursue his own research interests.[55] Seeman later abandoned the initial project and instead used the money to buy and improve a special X-ray spectrograph.[56]

Walter Steubing, *Privatdozent* of physics at the Technische Hochschule (technical university) of Aachen, needed special equipment to broaden his prewar investigation of the attenuation of the fluorescence of iodine vapor in a magnetic field. In his application he mentioned his recent paper on the subject, in which he argued that the phenomenon was not connected to the Zeeman effect and suggested that it could be due to an attenuation of the electron oscillations, caused by the magnetic field. He wished to investigate the influence of magnetic fields on the oscillation in the spectra of other substances.[57] Georg Wendt, too, assistant at the Physics Institute of the same Technische Hochschule, wished to continue his prewar investigation on the influence of an electric field on the triplet series of mercury and the doublet series of aluminium in order to contribute to the theoretical explanation of the Stark effect. Furthermore, he wanted to extend his research on the canal rays of carbon, silicon, and boron to other substances such as magnesium, calcium, and strontium in order to advance atomic theory and knowledge of the light emission mechanism.[58] Given that Steubing and Wendt belonged to the same institute and both needed instruments for spectroscopic analysis, the KWI für Physik allocated 10,000 marks to be shared between them, but at first only 2,050 marks were disbursed for the objective of a spectrograph.[59] For many years Steubing had been a close collaborator of Stark in several important investigations, for example, on the Stark effect and on the Doppler effect. Wendt had also been assistant of Stark, with whom he published several papers on canal rays and the Stark effect (Mehra and Rechenberg 1982, 103, vol. 1; Weinmeister 1925–1926, 1196–1198, vol. 2, s.v. "Stark, Johannes"). A few months after the grant was approved, Wendt left the institute and was replaced by Heinrich Kirschbaum, also a former private assistant of Stark, who then worked with Steubing on the behavior

---

[54]Einstein to Kuratorium, 25 April 1919, AMPG, I. Abt., Rep. 1 A, Nr. 1656, pp. 55–56; Einstein to Seemann, 28 April 1919, AMPG, I. Abt., Rep. 34, Nr. 10, folder Seemann; Einstein (2004, 560, 561).

[55]Seemann to Einstein, 11 May 1919 (Einstein 2004, 60–63, doc. 38).

[56]Seemann to Einstein, 2 May 1919 and 12 July 1920, AMPG, I. Abt., Rep. 34, Nr. 10, folder Seemann; (Seemann 1921).

[57]Steubing to Einstein, 6 April 1919, AMPG, I. Abt., Rep. 34, Nr. 10, folder Steubing; (Steubing 1919b). See also Steubing (1919a).

[58]Wendt to Einstein, 14 April 1919, AMPG, I. Abt., Rep. 34, Nr. 11, folder Wendt; (Wendt and Wetzel 1916; Wendt 1917).

[59]Einstein to Kuratorium, 25 April 1919, AMPG, I. Abt., Rep. 1 A, Nr. 1656, pp. 55–56; (Einstein 2004, 560); Einstein to Steubing, 27 May 1919, AMPG, I. Abt., Rep. 34, Nr. 10, folder Steubing; Einstein to Wendt, 27 May 1919, AMPG, I. Abt., Rep. 34, Nr. 11, folder Wendt; Einstein to KWG, 9 September 1919, AMPG, I. Abt., Rep. 1 A, Nr. 1657, p. 91.

of the nitrogen spectral lines in an electric field.[60] As Steubing explained in a later paper, this work was aimed not only at the systematic collection of data on the Stark effect but also at a better understanding of the atomic structure (Steubing 1922, 427). In early 1920, the KWI für Physik disbursed a further 500 marks for this research.[61]

Two applicants suggested the spectroscopic investigation of properties of materials. Friedrich Krüger, professor of physics and director of the Physics Institute at the Technische Hochschule of Danzig, applied for a fellowship for his assistant Helge Bohlin. Bohlin had developed a technique for X-ray spectroscopy which improved on the Debye-Scherrer powder method. The fellowship would allow him to continue the research on the crystallographic structure of metals and metal alloys that he had been performing for a year with this new promising method. As in the case of Seeman, the Direktorium decided to grant him 2,000 marks to buy instruments rather than granting a fellowship. Since Krüger had already bought the instruments he needed with the support of another foundation, he refused the money for the equipment and suggested to Einstein that he could hand over the instruments to the KWI für Physik in exchange for a fellowship of the same amount. Einstein, probably already annoyed by this kind of bureaucratic correspondence, took note of the refusal and did not respond to Krüger's suggestion. In the end, Krüger did not receive any money.[62] From the same Physics Institute of the Technische Hochschule of Danzig came the application of Karl Försterling, *Privatdozent* of theoretical physics, who received 2,000 marks for a project concerning the temperature dependency of the indices of refraction and absorption of metals in the infrared part of the spectrum. Försterling planned to extend to high temperatures the systematic measurements taken at low temperatures by Laue and Friedrich Franz Martens, adopting a method similar to theirs. The interesting theoretical rationale of his research was that "from the dependency of n [refractive index] and k [absorption index] upon temperature it would be possible to deduce the corresponding dependence of the number of electrons in the unit volume."[63]

Rudolf Seeliger, extraordinary professor of theoretical physics at the University of Greifswald, had in mind a broad research program on light emission by atoms, in particular he planned to work on "light emission in cathode dark space," "light emission in dissociated gases," "dependency of the spectrum of emitted light on the electron velocity," "determination of the critical excitation potential of selected

---

[60]Steubing to Einstein, 19 September 1919 and 30 January 1920, AMPG, I. Abt., Rep. 34, Nr. 10, folder Steubing; Wendt to Einstein, 2 October 1919, AMPG, I. Abt., Rep. 34, Nr. 11, folder Wendt; Einstein to Steubing, 11 October 1919, AMPG, I. Abt., Rep. 34, Nr. 10, folder Steubing.

[61]Einstein to KWG, 12 March 1920, AMPG, I. Abt., Rep. 1 A, Nr. 1657, p. 122.

[62]Krüger to Direktorium, 31 March 1919, AMPG, I. Abt., Rep. 34, Nr. 7, folder Krüger; Einstein to Kuratorium, 25 April 1919, AMPG, I. Abt., Rep. 1 A, Nr. 1656, pp. 55–56; Krüger to Einstein, 25 May, 2 June, and 18 August 1919, AMPG, I. Abt., Rep. 34, Nr. 7, folder Krüger; Einstein to Krüger, 25 May 1919, ibid.; Krüger to Laue, 8 February 1924, ibid. On Bohlin's new technique, see Bohlin (1920).

[63]Försterling to Einstein, 8 April 1919, AMPG, I. Abt., Rep. 34, Nr. 2, folder Försterling; Einstein to Kuratorium, 25 April 1919, AMPG, I. Abt., Rep. 1 A, Nr. 1656, pp. 55–56. On the method, see Laue and Martens (1907).

lines," and "light emission in proximity of the anode of a glow current." He had already obtained the support of other donors to buy the necessary equipment, and needed only the mercury to fill a Gaede pump.[64] Seeliger was a recognized expert on glow discharges in gases. An experiment he had performed with Ernst Gehrcke some years earlier had shown that "the emission of light of determined wavelengths is tied to a determined threshold electron velocity," that is, as Franck and Gustav Hertz later proved, to the exchange of an energy quantum.[65] The KWI fulfilled Seeliger's moderate request with 1,000 marks for which it was thanked in a paper on the excitation conditions of mercury spectral lines.[66]

In his letter asking for support for the investigation of "photoelectric effect and X-rays," Robert W. Pohl, extraordinary professor of experimental physics at the University of Göttingen, did not even bother to explain his work in detail but took it for granted that Einstein already knew about it. Indeed, Pohl was well known to the Direktorium members since he had been a student of Warburg and worked until 1916 as *Privatdozent* at the Physical Institute of the Berlin University. Furthermore, he had already worked with Peter Pringsheim on the photoelectric properties of alkaline metals and published a book on this subject. Not surprisingly, Pohl was given 5,000 marks for unspecified "material and instruments for scientific research."[67] The major focus of Pohl's research became at that time the effects of light irradation on the electrical conductivity of solids, in particular of crystals, on which he and his student (and later assistant) Bernhard Gudden worked in the following years with repeated support from the KWI für Physik.[68]

Two applications asked for support for measurements concerning the electrical properties of different materials. Wilhelm Hammer, assistant at the University of Freiburg im Breisgau, had developed a new method for measuring capacitances and constants of dielectricity by means of electric oscillations, which yielded results that were three times more precise than before. This improvement would allow the solution of several pending questions which were mentioned in a letter of support by Hammer's superior, the ordinary professor of physics Franz Himstedt. He and Hammer intended to test, for example, the Clausius-Mosotti relation in gases and liquids at different pressures, as well as a hypothesis by Debye concerning pre-existing dipoles in the material in contrast to the dipole moments induced by the electro-

---

[64] Seeliger to Einstein, 31 March 1919, AMPG, I. Abt., Rep. 34, Nr. 10, folder Seeliger.

[65] Hoffmann (1986, 279); see also Wilhelm (1987).

[66] Einstein to Kuratorium, 25 April and 2 May 1919, AMPG, I. Abt., Rep. 1 A, Nr. 1656, pp. 55–56 and 57 respectively; Seeliger to Einstein, 29 April 1919, AMPG, I. Abt., Rep. 34, Nr. 10, folder Seeliger; (Seeliger 1920).

[67] Einstein to Kuratorium, 25 April 1919, AMPG, I. Abt., Rep. 1 A, Nr. 1656, pp. 55–56; see also Pohl to Einstein, 16 April 1919, AMPG, I. Abt., Rep. 34, Nr. 9, folder Pohl; (Pohl and Pringsheim 1914).

[68] For a bibliography of Pohl's and Gudden's scientific papers on photoelectric conductivity and other effects of light irradiation, in many of which the authors thanked the KWI für Physik for financial support, see s.v. "Pohl, Robert" and "Gudden, Bernhard" in Weinmeister (1925–1926) and in Stobbe (1936–1940). See also Gudden (1944, 167), Gerlach (1978, 217–218), Hund et al. (1988, 189–190).

magnetic field. Hammer was given 5,000 marks for "instruments for measurement of electric oscillations."[69] Walther Kaufmann, professor of experimental physics at the University of Königsberg, received 3,000 marks for equipment to generate short-wave electromagnetic signals, with which he intended to perform vacuum and magnetic measurements in high-frequency fields. In particular, he wanted to measure the frequency dependence of, again, the constants of dielectricity and absorption at high frequencies. The support of the KWI für Physik was later acknowledged in two papers reporting measurements of pressure in vacua by means of ionization currents of hot cathodes, and of the reversible magnetic permeability of iron at high frequencies.[70] More than ten years earlier, Kaufmann had done pioneering work on the velocity dependence of the inertial mass of electrons. His measurements seemed to contradict special relativity theory, and both Einstein and Planck disputed their theoretical implications.[71]

An application concerned with a much-debated problem came from the professor of physics at Berlin's Landwirtschaftliche Hochschule (Agriculture College), Erich Regener,[72] whom Einstein, some years earlier, had lauded as one of "the best young physicists in Germany" along with Johann Koenigsberger and Edgar Meyer.[73] Although often labeled a measurement of the "elementary quantum" at the time, Regener's investigation had nothing to do with Planck's quantum. Remember that in 1910–1911 Felix Ehrenhaft and his collaborators at the University of Vienna had claimed to have measured electric charges with different values on small particles of different materials, including values much lower than the charge of an electron, and therefore assumed the existence of "subelectrons." On the other hand, in those same years Robert A. Millikan of the University of Chicago was improving his measurement of the electric charge on individual droplets and collecting considerable evidence for the existence of a definite elementary charge. This became a controversial question, not yet resolved in the late twenties.[74] There were more disbelievers than adherents to Ehrenhaft's results—which Einstein, too, considered "illusory."[75] Besides its strictly-speaking physical aspects, the controversy was also, indirectly, about the underlying assumptions concerning the constitution of matter and electricity, namely atomism or continuum theory (Holton 1978, 222). Planck, Warburg, Rubens as well as many others took part in the debate at congresses or published data

---

[69] Einstein to Kuratorium, 25 April 1919, AMPG, I. Abt., Rep. 1 A, Nr. 1656, pp. 55–56; Hammer to Einstein, 27 March 1919, AMPG, I. Abt., Rep. 34, Nr. 4, folder Hammer; Himstedt to Einstein, 27 March 1919, ibid., folder Himstedt. On the new method, see Hammer (1919–1920). We could not find any report on the envisaged measurements.

[70] Kaufmann to KWI für Physik, 8 April 1919, AMPG, I. Abt., Rep. 34, Nr. 6, folder Kaufmann; Einstein to Kuratorium, 25 April 1919, AMPG, I. Abt., Rep. 1 A, Nr. 1656, pp. 55–56; (Kaufmann and Serowy 1921; Urbschat 1921).

[71] Einstein (1989, 270–272) editorial comment "Einstein on the theory of relativity."

[72] Regener to Einstein, 14 April 1919, AMPG, I. Abt., Rep. 34, Nr. 9, folder Regener.

[73] Einstein to Laub, 10 August 1911, in Einstein (1993b, 308, doc. 275).

[74] On the Millikan-Ehrenhaft dispute, see Holton (1978).

[75] Quotation from "Discussion du rapport de M. Perrin" in Langevin and de Broglie (1912, 251), Einstein (1993a, 509, doc. 25).

of their own. Regener too had, in 1911, arrived at results supporting Millikan's view on the definite value of the smallest electric charge (Holton 1978, 200, note 115). In his application, he now proposed to look closely at the statistics of the methods used for measuring the elementary charge and further to determine it afresh by counting and by measuring charge in canal rays. In May 1919—that is, after the Direktorium meeting of 24 April at which the grants had been approved—Ehrenhaft also inquired about funds from the KWI für Physik, remarking that Einstein was well aware of his research.[76] A response from Einstein is not documented, but Ehrenhaft did not receive money, whereas Regener received 7,000 marks with which he bought several instruments.[77]

Regener's is one of the few research projects supported by the KWI für Physik about which a scientific comment can be found in Einstein's correspondence. Probably at the beginning of June 1920, Einstein wrote to Paul Ehrenfest:

> Moreover, under Regener an investigation concerning the Ehrenhaft question has been made by use of droplets; it shows that the apparently lower values of the elementary quantum are produced by mobilities that are too small; these are, probably, caused by gas layers increasing the hydrodynamically effective radius of the droplet.[78]

Regener's paper expounding the conclusions mentioned by Einstein in his letter had just been presented on 3 June 1920 at a meeting of the Academy of Science (Regener 1920), while the results of the measurements taken by Regener's co-worker Ernst Radel appeared a few weeks later (Radel 1920).

The only research on thermodynamics supported was proposed by Alfred Magnus, *Privatdozent* of physical chemistry at the University of Tübingen, who wanted to continue his series of measurements of specific heat coefficients of solids for high temperatures, in particular carbon, silicon, and silicon carbide. In order to improve the precision of his measurements he needed different kinds of thermometers, for which he received 3,000 marks. Unfortunately, a year later Magnus reported that he could not buy all the instruments needed due to inflation and could not complete the planned measurements because of teaching duties. Very correctly, he returned some of the money.[79]

Finally, one project concerning classical kinetic theory was sponsored: Wilhelm Westphal, extraordinary professor of physics at the University of Berlin and former student of Rubens, received 3,000 marks for the continuation of experimental work aimed at "shedding light upon the little-known theory of the radiometer at higher pressure and on Knudsen's accommodation coefficient."[80] It seems that Einstein played

---

[76]Ehrenhaft to Einstein, 28 May 1919 (Einstein 2004, 73, doc. 46). The previous year Einstein and Ehrenhaft had corresponded on the subject (Einstein 1998, 861–862, 902–905, docs. 605, 630).

[77]Einstein to Kuratorium, 25 April 1919, AMPG, I. Abt., Rep. 1 A, Nr. 1656, pp. 55–56; Regener to Einstein, 12 June 1919, AMPG, I. Abt., Rep. 34, Nr. 9, folder Regener.

[78]Einstein to Ehrenfest, undated [6 June 1920], in Einstein (2006, 297, doc. 46).

[79]Magnus to Einstein, 7 April 1919 and 11 July 1920, AMPG, I. Abt., Rep. 34, Nr. 8, folder Magnus; Einstein to Kuratorium, 25 April 1919, AMPG, I. Abt., Rep. 1 A, Nr. 1656, pp. 55–56.

[80]Westphal to Einstein, 16 April 1919, AMPG, I. Abt., Rep. 34, Nr. 11, folder Westphal; Einstein to Kuratorium, 25 April 1919, AMPG, I. Abt., Rep. 1 A, Nr. 1656, pp. 55–56.

a role, at least indirectly, in Westphal's initiative. In February 1919 Westphal had submitted to the journal *Verhandlungen der Deutschen Physikalischen Gesellschaft* a paper expounding a theory of the radiometer at high gas pressure, with a formula for the momentum of gas molecules hitting the radiometer (Westphal 1919a). The theory was said to be in accordance with experimental data on which Westphal lectured a few weeks later, at a meeting of the Deutsche Physikalische Gesellschaft held on 14 March 1919 under Einstein's chairmanship.[81] On that occasion Westphal again expounded his "attempt at a theoretical interpretation," thus probably eliciting some reflections from Einstein. In fact, in a short note retracting his previous theory while announcing a new systematic investigation, Westphal declared some months later: "Mr. Einstein in particular was so kind as to make me aware of a contradiction concerning the momentum formula" (Westphal 1919b). Westphal had based his formula on the inadmissible assumption of a uniform molecular velocity in small spaces. In three subsequent papers, in which he duly thanked the KWI für Physik for its support, Westphal dealt with the dependency of the radiometric effect upon gas pressure, but no longer attempted to develop a theoretical interpretation (Westphal 1920a, b, 1921).

With respect to the original motivation for the foundation of the institute, we found it particularly odd that in January 1920 some of the funds were also spent to support the journal *Physikalische Berichte*.[82] In the winter of 1919 an ad hoc committee of the Deutsche Physikalische Gesellschaft, of which Haber and Einstein among others were members, brought about a radical change in the editorial policy concerning the physics journals. Due to the difficult financial situation of the association, the scope of its free publication, the *Verhandlungen*, was considerably reduced and instead the new subscription journal *Zeitschrift für Physik* was published starting in January 1920. At the same time, both the Deutsche Physikalische Gesellschaft and the newly established Deutsche Gesellschaft für technische Physik (German Society of Technical Physics) launched the *Physikalische Berichte* as a journal of abstracts on both German and foreign physics literature, pure and applied.[83] The KWI für Physik and the Deutsche Physikalische Gesellschaft gave 5,000 marks each for this new journal, while the Prussian Academy contributed 10,000 marks in compliance with an application submitted by Einstein, Nernst, Planck, Rubens, and Warburg.[84]

---

[81]"[Bericht über die] Sitzung vom 14. März 1919" *Verhandlungen der Deutschen Physikalischen Gesellschaft* 21 (1919, 69); Westphal and Gerlach (1919).

[82]Einstein to Kuratorium, 28 January 1920, AMPG, I. Abt., Rep. 1 A, Nr. 1657, p. 114.

[83]"Neuordnung des Zeitschriftenwesens der Deutschen Physikalischen Gesellschaft," *Verhandlungen der Deutschen Physikalischen Gesellschaft*, 21 (1919), pp. 673–675. On the editorial changes in the physics press, see Forman (1968, 171–205), Dreisigacker and Rechenberg (1995, F136–F138), Hermann (1995, F79–F80).

[84]Minutes of the meeting of the physical-mathematical class, 23 October 1919, AAdW, II–V, Bd. 134, p. 112; (Kirsten and Treder 1979, 124–125). The grant was approved by the appropriation committee of the Academy on 6 November 1919 (Kirsten and Treder 1979, 235, vol. 2). The main support amounting to 65,000 marks came from the Deutsche Gesellschaft für technische Physik (Hoffmann and Swinne 1994, 37, 48–49).

Rejected Proposals

Some projects were not funded because they did not belong to physics proper, at least not in the opinion of the Direktorium: Himstedt, who was mentioned above as supporting Hammer's application, had asked for instruments for meteorological measurements.[85] Hans Rosenberg, professor of astronomy at the University of Tübingen, wished to apply a new photoelectrical method to astronomical measurements.[86] Hermann Starke, professor of physics at the Technische Hochschule of Aachen, needed new costly instruments to continue his investigation on electric resonance, relaxation effects, and distortions in current-voltage curves at high frequencies.[87] Einstein responded that the institute had decided to support only "purely scientific (not more or less technical)" research.[88]

An interesting project, which was probably rejected for scientific reasons, was suggested by Otto Lehmann, professor of physics at the Technische Hochschule of Karlsruhe. Lehmann proposed studying the molecular directional forces exerted on liquid crystals by a magnetic field. He was nearing retirement, but intended to continue his investigations privately.[89] Lehmann is regarded as the discoverer of the liquid crystals on which he had worked almost exclusively for thirty years. Although his vitalistic explanations of the phenomena were disputed, his empirical studies on the subject were highly esteemed (Schleiermacher and Schachenmeier 1923; Brauns 1934). His "famous publications" had even been cited by Jean Perrin in his talk at the First Solvay Congress in 1911 (Perrin 1912, 227). Nevertheless, Lehmann's pending retirement was not the only reason for the refusal, as Einstein, in his letter turning down the proposal, not only deplored the lack of funds but also admitted that he had not understood Lehmann's "remark on the magnetic-electric induction" and commented that "the molecular structure is not essential for the theoretical interpretation of these phenomena."[90]

Finally, two applications were rejected because they asked for fellowships. While it is understandable that Einstein refused to support a student who wanted to develop "military-technical inventions,"[91] it is highly surprising that the application of Wilhelm Lenz, assistant to Sommerfeld at the University of Munich, was also rejected. Lenz wanted "to take up again the theory of mono-atomic gases under the point

---

[85]Himstedt to Einstein, 27 March 1919, and Einstein to Himstedt, 26 April 1919, AMPG, I. Abt., Rep. 34, Nr. 4, folder Himstedt.

[86]Rosenberg to Einstein, 11 April 1919, and Einstein to Rosenberg, 26 April 1919, AMPG, I. Abt., Rep. 34, Nr. 9, folder Rosenberg.

[87]Starke to Einstein, 8 April 1919, AMPG, I. Abt., Rep. 34, Nr. 10, folder Starke.

[88]Einstein to Starke, 26 April 1919, AMPG, I. Abt., Rep. 34, Nr. 10, folder Starke.

[89]Lehmann to Einstein, 26 March, 13 April, 16 April 1919, AMPG, I. Abt., Rep. 34, Nr. 8, folder Lehmann.

[90]Einstein to Lehmann, 28 April 1919, AMPG, I. Abt., Rep. 34, Nr. 8, folder Lehmann; (Einstein 2004, 562).

[91]Georg Krakow to Einstein, undated [March 1919] and 1 May 1919, and Einstein to Krakow, 6 May 1919, AMPG, I. Abt., Rep. 34, Nr. 7, folder Krakow.

of view of the quantization of the collision processes."[92] In this context, that is, in dealing with the quantum theory of non-periodic processes, he intended to look at the nature of continuously distributed X-rays. Lenz also envisaged investigating the nuclear structure from the angle of quantum theory. Despite the recommendation for Lenz by Sommerfeld, and his remarks that Sommerfeld had "discussed the matter a year ago with Einstein and Planck," Einstein answered that the KWI had decided not to award fellowships and did not even offer Lenz money to buy instruments.[93] As a reason for the refusal Einstein referred to the difficult financial situation of the university institutes and to the large number of applications received,[94] as if fellowships claimed an inordinate part of the funds and thus hindered a broader distribution. Yet Einstein's justifications are not convincing, not only because in the cases of Bohlin and Seemann the KWI allocated money for instruments instead of the requested fellowships, but also because at the end of September 1919, when all the grants approved in April had been paid, the institute still had about 65,000 marks left to distribute.[95] For whatever reason, the leaders of the institute did not want to support research projects under Sommerfeld's supervision.

Negative Reactions to the Institute's Policy. Annual Report 1919/20

The distribution policy of the Direktorium did not meet with the unconditional approval of all members of the KWI für Physik. Commenting on Einstein's communication about the grants approved on 24 April 1919, the chairman of the KWI für Physik, Siemens, complained that, unlike the procedure of the other KW Institutes, no discussion on the research projects had taken place in the KWI für Physik and that a report on the institute's activities for the fiscal year 1918/19 had not yet been delivered. He also criticized the budgetary plan for 1919/20.[96] As for Krüss, the representative of the Prussian Ministry of Education, he criticized the fact that "the distribution of the funds was based on the criterion of the needs of the university institutes." For reasons of national prestige, he would have preferred a concentration of the available money "in order to possibly also engage one day in a larger endeavor (measurement of radiation in another climate or the like)." However, he gave his approval "because, apparently, the majority of the projects approved are directed at the same field of radiation theory."[97] The three scientific members of the KWI für Physik obviously had no objections since they were also members of the Direktorium. The decisions of the Direktorium were approved almost two months later, on

---

[92]Lenz to Einstein, with a postscript by Sommerfeld, 25 March 1919, AMPG, I. Abt., Rep. 34, Nr. 8, folder Lenz; (Einstein 2004, 18–19, doc. 11). See also Sommerfeld to Einstein, 25 March 1919, in Einstein (2004, 20, doc. 12).

[93]Einstein to Lenz, 26 April 1919, AMPG, I. Abt., Rep. 34, Nr. 8, folder Lenz.

[94]See also Einstein to Krüger, 25 May 1919, AMPG, I. Abt., Rep. 34, Nr. 7, folder Krüger, and Einstein to Seemann, 28 April 1919, AMPG, I. Abt., Rep. 34, Nr. 10, folder Seemann.

[95]Bank statement 1 April–31 December 1919, AMPG, I. Abt., Rep. 34, Nr. 8, folder Mendelssohn.

[96]Von Siemens to Glum, 14 May 1919, AMPG, I. Abt., Rep. 1 A, Nr. 1665, pp. 27–30; see also von Siemens to Glum, 2 May 1919, ibid., p. 24.

[97]Krüss to von Siemens, 5 June 1919, addition to the circular letter of von Siemens to the Kuratorium members, 2 May 1919, AMPG, I. Abt., Rep. 1 A, Nr. 1656, p. 63.

16 June 1919, much to the annoyance of Einstein, who meanwhile had complained twice about the delay and also had to provide further explanations concerning the plan for 1919/20.[98]

In the report on the institute's activities for the period April 1919 to March 1920 it is clear that the original idea had been diluted:

> The task of the Kaiser-Wilhelm-Institut für Physik is primarily the support of large-scale scientific research projects which cannot be carried out with the means of the single [university] institutes. However, the difficult economical situation and the scarcity of funds for research of the various institutes resulting therefrom made it necessary to stray somewhat away from this plan and to support a larger number of separate research projects. The Direktorium believed that in this way it could best serve the continuation and improvement of physical research.[99]

The report then mentions that a group of fifteen projects had been funded, "a large part of which was devoted to spectroscopy," and goes on to describe in some further detail the work of Freundlich in testing general relativity.

In summary, we may say that most of the projects were indeed concerned with spectroscopy and radiation phenomena, in which the new concept of Planck's quantum played a dominant role. Others dealt with properties of material systems and with molecular physics. A number of subfields of physics were not present at all, for example, hydrodynamics, acoustics, or pure optics. One possible explanation for the concentration of research projects in the fields mentioned could be that these were becoming subjects of increasing interest among physicists; this was not an entirely new trend though, since in many cases the proposals aimed at the continuation of research programs that had already started before the war. It must also be considered that spectroscopic research probably required the purchase of expensive equipment, for which the university institutes lacked money. This would explain the greater need to apply for outside support as compared to research in other fields.

In any case, only the testing of Planck's formula and the research sponsored by Nernst and Einstein had been directly initiated by Direktorium members. Thus only the testing of Planck's formula can be regarded as a step in a research strategy directed toward quantum physics, and this was limited to measurements and did not address new ideas. The other two initiatives corresponded to the particular interests of their initiators at the time. Nernst abandoned his research on gas degeneracy in the following years. As for Einstein, it is quite clear that he was utilizing the institute's funds solely for research on relativity theory. Einstein was certainly also interested in the testing of Planck's law and in other problems of quantum physics, but not so interested as to initiate research in this field. On the other hand, the other Direktorium members seemed unwilling to support research on relativity theory with

---

[98]Einstein to Kuratorium, 7 May and 16 June 1919, AMPG, I. Abt., Rep. 1 A, Nr. 1656, pp. 59 and 64 respectively; von Siemens to Einstein, 16 June 1919, AMPG, I. Abt., Rep. 34, Nr. 13; Einstein, "Erläuterungen zu dem Haushaltsplan," undated [sent to von Siemens on 9 May 1919], AMPG, I. Abt., Rep. 34, Nr. 13.

[99]Report of the KWI für Physik 1 April 1919–31 March 1920, AMPG, I. Abt., Rep. 1 A, Nr. 1665, p. 48; (Kirsten and Treder 1979, 152–153, vol. 1). The report was sent to the Kuratorium in September 1920 (Einstein to Schmidt-Ott, 13 September 1920, AMPG, I. Abt., Rep. 1 A, Nr. 1665, p. 45).

all the necessary funds. Einstein had to turn to a private donor to support Grebe and Bachem even though the institute still had funds available.

In fact, the KWI was intentionally holding back its funds. As Einstein declared in the explanations relating to the plan for 1919/20, the Direktorium had decided to distribute only the new income while the previous surplus was to be kept in reserve.[100] Therefore, during that fiscal year the institute spent only 76,559 marks, 66,050 of which were allocated for "scientific purposes." At the end of the period the bank statement still showed a surplus of 83,926 marks, that is, more than the amount handed out during the past twelve months.[101]

3.3   The Period 1920/21

The applications for equipment to be provided by the KWI für Physik kept coming in at a steady but not growing rate. Fourteen applications were submitted from May 1919 to April 1920, a further seven came in before the end of November 1920. The Direktorium did not meet until 22 April 1920 so that some applicants—for example, Wilhelm Hallwachs and Peter Paul Koch—had to wait for almost a year before receiving an answer. At that meeting, six research projects were approved. Seven further projects were approved at the Direktorium meetings of 8 July and 2 December 1920. Five applications were rejected for various reasons.[102] In one case, that of Reinhold Fürth, the decision was postponed until more precise explanations could be submitted. For four inquiries there is no record of any official decision on the part of the Direktorium.

Old Projects and New Proposals

Again, the majority of approved applications concerned spectroscopy, light emission from atoms through the inelastic scattering of electrons, the photoelectric effect, and the elementary electric charge. Some of the allocations allowed the continuation of projects already supported. Continuing his work on testing the theory of relativity, Freundlich, in Potsdam, investigated the cyanogen bands with a spectroscopic oven endowed not by the KWI für Physik but by Carl Bosch, the managing director of the Badische Anilin- und Sodafabrik (B.A.S.F.).[103] Freundlich also worked on stellar statistics and, "in collaboration with Einstein," on different methods of determining

---

[100]Einstein, "Erläuterungen zu dem Haushaltsplan," undated [sent to von Siemens on 9 May 1919], AMPG, I. Abt., Rep. 34, Nr. 13.

[101]Bank statement 1 January–31 March 1920, AMPG, I. Abt., Rep. 34, Nr. 8, folder Mendelssohn; survey of income and expenditure 1 April 1919–1931 March 1920, AMPG, I. Abt., Rep. 1 A, Nr. 1665, pp. 46–47. See also Tables 3.1 and 3.2. The new balance given in Table 3.1 includes the cash.

[102]Einstein to Kuratorium, 30 April, 3 May, 14 September and 7 December 1920, AMPG, I. Abt., Rep. 1 A, Nr. 1657, pp. 125–126, 146, 151; (Kirsten and Treder 1979, 151–153, vol. 1; Kant 1987, 133).

[103]Freundlich to Einstein, 12 August 1920, in Einstein (2006, 371–372, doc. 101); see also Freundlich to von Harnack, 15 December 1919, AMPG, I. Abt., Rep. 1 A, Nr. 1657, pp. 106–107.

stellar masses (Müller 1921, 118–119). Steubing received a further 6,300 marks, from the 10,000 assigned in 1919, to continue the investigation of nitrogen spectral lines in an electric field conducted in collaboration with Kirschbaum, who in a later paper acknowledged the support of the KWI. With the apparatus financed by the KWI Steubing also worked on the influence of temperature on the iodine spectral band.[104] Similarly, Wagner's X-ray spectroscopy was supported by an additional 1,500 marks. Both Wagner and his student, Helmuth Kulenkampff, later thanked the KWI in papers on the relation between wavelength and the intensity of X-ray reflection from crystals, and on the continuous X-ray spectrum.[105] Pohl, who had meanwhile become a full professor and director of one of the two institutes for experimental physics at the University of Göttingen, obtained 8,000 marks to continue with his assistant, Gudden, the investigation of the photoelectric effect. By studying the increase of the dielectric constant of solids due to light irradiation, they had discovered that the latter elicits eigenvibrations of electrons in the solid's interior. In order to establish whether this was a general effect, they needed new instruments to extend the investigation to diamonds and other crystals.[106] Seeliger, too, worked on the same kind of problems as before, namely on the emission of light through the collision of electrons with hydrogen atoms, and recently also with helium atoms. He was granted an additional 2,000 marks to buy a set of high-voltage accumulators.[107] The support, continued in the following years, was acknowledged in several articles.[108] Another 5,000 marks went to Otto von Baeyer for the continuation of Regener's and Radel's work concerning the determination of the elementary electric charge. Baeyer was to become the new director of the Physical Institute of the Landwirtschaftliche Hochschule in Berlin, replacing Regener, who had moved to Stuttgart. Von Baeyer wanted to bring to an end the series of measurements on droplets, under different high pressures and with the "Einstein-Weiss method."[109] The work, done by doctoral candidate Kurt Wolter, led to the confirmation of Millikan's and Regener's results but also showed the "inaccuracy" of the "Einstein-Weiss method" (Wolter 1921, 351).

The question of the elementary electric charge intrigued others as well. We have already mentioned the request made by Ehrenhaft in May 1919, which should have

---

[104]Einstein to Bank Mendelssohn, 27 July 1920, AMPG, I. Abt., Rep. 1 A, Nr. 1657, p. 144; (Kirschbaum 1923; Steubing 1921).

[105]Wagner to Einstein, 18 June and 14 August 1920, and Einstein to Wagner, 1 July 1920, all in AMPG, I. Abt., Rep. 34, Nr. 11, folder Wagner; Einstein to Kuratorium, 7 December 1920, AMPG, I. Abt., Rep. 1 A, Nr. 1657, p. 151; (Wagner and Kulenampff 1922; Kulenkampff 1922).

[106]Pohl to "Geheimrat," 3 July 1920, and to Einstein, 14 July 1920, AMPG, I. Abt., Rep. 34, Nr. 9, folder Pohl; Einstein to Pohl, 10 July 1920, ibid.; Einstein to Kuratorium, 7 December 1920, AMPG, I. Abt., Rep. 1 A, Nr. 1657, p. 151.

[107]Seeliger to Einstein, undated [January 1920] and 2 May 1920, and Einstein to Seeliger, 30 April 1920, all in AMPG, I. Abt., Rep. 34, Nr. 10, folder Seeliger; Einstein to Kuratorium, 3 May 1920, AMPG, I. Abt., Rep. 1 A, Nr. 1657, p. 126; (Kirsten and Treder 1979, 152, vol. 1).

[108]See, e.g., Seeliger and Mierdel (1921), Seeliger and Thaer (1921), Seeliger (1922a, b).

[109]Von Baeyer to Einstein, 30 June and 15 November 1920, and Einstein to O. von Baeyer, 30 July 1920, all in AMPG, I. Abt., Rep. 34, Nr. 1, folder Baeyer; Einstein to Kuratorium, 7 December 1920, AMPG, I. Abt., Rep. 1 A, Nr. 1657, p. 151.

been dealt with at a Direktorium meeting of the current fiscal year 1920/21. Unfortunately there is neither mention of such a discussion in the minutes of the meetings nor an official reply. But the Direktorium certainly discussed the question in relation to the proposal submitted in October 1920 by Fürth, *Privatdozent* of physics at the German University of Prague, in the same period as the support for measurements taken under von Baeyer's supervision was approved. Until then Fürth, a student of Philipp Frank and Anton Lampa, had worked in particular on Brownian motion and had published many papers on the subject.[110] In the context of his research on microscopic particles, he had come to doubt the evidence for the elementary electric charge for reasons relating, among other things, to the method of measurement and to the statistics of the results (Fürth 1920). According to him, the dynamics of small moving particles, in the usual experimental procedure according to Millikan's and Ehrenhaft's methods, was too complicated. He now wanted to determine the elementary electric charge with a torsion balance of the Eötvös type in a static setup, that is, without measuring the mass of the charge carriers. In the application Fürth enclosed a sketc.h of a "torsion balance for the measurement of the smallest charges" and explained how it worked.[111] Einstein replied that the Direktorium "could not yet come to a decision for giving you the means requested in your communication, because it had doubts as to whether your plan could be carried out" and asked for further information with which the experiment's feasibility could be assessed. As a postscript he added: "I admit that I myself have not supported your proposal because I believe that the expected success of the investigation will be small, in comparison with the difficulties to be overcome."[112] Fürth and Einstein must then have met during the latter's visit to Prague in January 1921 and discussed the matter because, in his renewed application to the KWI für Physik, Fürth refers to the "verbal agreement" reached by them.[113] His request of 2,000 marks was finally approved in the following budget year.[114] Einstein's initial skepticism was justified: As late as February 1926 1926 Fürth's measurements had not yet come to an end.[115]

New projects addressing questions of quantum physics were also proposed. In one particular case, the Direktorium showed considerable interest, approving a project even before the application was submitted. On 8 July 1920 10,000 marks were granted to Gerhard Hettner, Rubens's assistant at the Physical Institute of the University of Berlin, for the purchase of instruments to be used for an "investigation into the structure of infrared gas spectra."[116] In his subsequent application Hettner explained that he wanted to find out "whether an electric field has any effect on the two [absorption] bands of gases [like HCl] in the infrared," adding that "the investigation would be of

---

[110]Later Fürth also edited a collection of Einstein's papers on Brownian motion (Einstein 1922a).

[111]Fürth to Einstein, 19 October 1920, AMPG, I. Abt., Rep. 34, Nr. 2, folder Fürth.

[112]Einstein to Fürth, 10 December 1920, ibid. (see footnote 111).

[113]Fürth to Einstein, 21 January 1921, AMPG, I. Abt., Rep. 34, Nr. 2, folder Fürth; see also Fürth to Einstein, 18 December 1920, ibid. On Einstein's visit to Prague, see Fölsing (1993, 560).

[114]Einstein to Kuratorium, 7 March 1921, AMPG, I. Abt., Rep. 1 A, Nr. 1658, pp. 166–167.

[115]Fürth to Laue, 12 July 1924 and 15 February 1926, AMPG, I. Abt., Rep. 34, Nr. 2, folder Fürth.

[116]Einstein to Kuratorium, 14 September 1920, AMPG, I. Abt., Rep. 1 A, Nr. 1657, p. 146.

great importance because Bohr's theory and Planck's second quantum theory lead to different effects, so that a decision between them would be possible."[117] In fact, in an enclosed paper Hettner calculates the effect expected according to Bohr's theory but defers to a later paper the calculation in accordance with Planck' second quantum theory (Hettner 1920a).

It is likely that Hettner received support even before he asked for it because his work had already been discussed in the Berlin physics community in the previous months and attracted the attention of the Direktorium members. In his correspondence Einstein himself expressed particular interest in the experimental research that had led Hettner to formulate the hypothesis mentioned in his application. In that work Hettner

> [...] has shown that the gas spectra in the infrared HCl and H_2O can be explained completely by (quasi)elastic eigenvibrations as well as by combined frequencies. If, e.g., ν_1 and ν_2 are absorption frequencies then $2\nu_1, 2\nu_2, \nu_1 + \nu_2$ also occur, as is to be expected by the quantum theory of imperfectly elastic vibrations.[118]

Unfortunately, Hettner did not carry out the planned research but instead shifted his interest to radiometer physics.[119] The promised paper, which was to handle infrared gas spectra in an electric field according to Planck's second quantum theory, never appeared. The theory was abandoned but not because of any contributions by Hettner.

Christian Füchtbauer, extraordinary professor at the University of Tübingen, received 7,000 marks to buy several special optical devices that would allow him to continue with higher precision a series of measurements of the intensity and width of spectral lines. He had developed a method for obtaining not just a few lines but the entire spectrum of an atom, by exposing mercury vapor to the light of an intense mercury lamp. With his measurements he wished to tackle several questions. First, he wanted to establish whether the frequency of electronic transitions depends upon the variations in the line width with the gas density. Second, he wanted to determine whether the Stark effect was a sufficient explanation of the line broadening.[120]

Koch had submitted his application in June 1919, immediately after being appointed professor of experimental physics at the Physikalisches Staatslaboratorium of Hamburg.[121] He wished to continue his investigations into "the intensity distribution in spectral lines" which, he explained, "are closely connected with theoretical problems resulting from Sommerfeld's extension of Bohr's [atomic] theory."

---

[117]Hettner to Einstein, 20 July 1920, AMPG, I. Abt., Rep. 34, Nr. 4, folder Hettner.

[118]Einstein to Ehrenfest, undated [June 1920] (Einstein 2006, 297, doc. 46). The paper referred to is Hettner (1920b). Einstein also mentioned Hettner's work in a letter to Hendrik Antoon Lorentz of the same period (Einstein 2006, 313).

[119]See s.v. "Hettner, Gerhard" in Stobbe (1936–1940, 1105, vol. 2).

[120]Füchtbauer to Kuratorium, 2 November 1920 and 31 January 1921, AMPG, I. Abt., Rep. 34, Nr. 2, folder Füchtbauer; Einstein to Kuratorium, 7 December 1920, AMPG, I. Abt., Rep. 1 A, Nr. 1657, p. 151; (Mehra and Rechenberg 1982, 653, vol. 1). For the results of Füchtbauer's work, see Fuchtbauer and Joos (1922), Fuchtbauer et al. (1923).

[121]The State Laboratory of Physics was about to become an institute of the newly established university (Fouquet 1999, 160–162).

In addition, he also envisaged conducting research on the effects of light on silver bromide.[122] Koch, a student of Röntgen, had worked for many years at the University of Munich, where Sommerfeld also taught. He was a specialist on the intensity of spectral lines and thus in photometry. He had previously developed a new spectrograph for mapping line profiles that measured the degree of blackening of photographic plates using photocells (Koch 1912). In 1919, Freundlich installed such an instrument at the Astrophysikalisches Observatorium in Potsdam, possibly for his daylight measurements of light deflection at the solar limb. The same spectrograph was also used by Grebe and Bachem for their measurements of the gravitational redshift of the cyanogen bands.[123] Only in April 1920 did the KWI für Physik give Koch the 12,800 marks he had requested, that is, the biggest sum distributed in this fiscal year.[124] Judging from his later papers though, Koch seems to have abandoned the first topic mentioned in his letter and concentrated his research almost exclusively on photographic processes.[125]

Franck, who in 1920 was still head of a department in Haber's KWI für physikalische Chemie und Elektrochemie in Berlin, first received 2,000 marks and later a further 10,000 for the continuation of his research on the collision of electrons with atoms.[126] As is well known, the experiments on the collision of electrons with gas atoms performed before the war by Franck and Hertz at Haber's institute had unwittingly led to the confirmation of the quantum relationship between energy and frequency and of the Bohr-Sommerfeld atomic theory (Mehra and Rechenberg 1982, 197–200). In his first application to the KWI für Physik, Franck asked for support for the improvement of a newly conceived photoelectric method of "fixing," that is, registering the spectral lines emitted by the collision of electrons with gas molecules. With this method it was possible to narrow the gap in the spectral range between optical and X-ray lines, which was inaccessible to experimentalists at the time. Recent experiments even "let it appear promising to apply this procedure to the impact of electrons on the surface of solids."[127] In a second application, some months later, Franck asked for a greater sum that would allow his assistant, Paul Knipping, to implement a method for

---

[122] Koch to Einstein, 8 June 1919, AMPG, I. Abt., Rep. 34, Nr. 7, folder Koch; see also Koch to Kuratorium, 24 May 1920, AMPG, I. Abt., Rep. 1 A, Nr. 1657, p. 134.

[123] "Bericht über die Tätigkeit im Jahre 1919," AMPG, I. Abt., Rep. 34, Nr. 2, folder Freundlich; Grebe to Einstein, 17 April 1919, in Einstein (2004, 37–38, doc. 25); Einstein to Grebe, 26 April 1919, AMPG, I. Abt., Rep. 34, Nr. 3, folder Grebe; (Hentschel 1992c, 24–26). Also Steubing and Füchtbauer used a Koch microphotometer (Steubing 1921; Füchtbauer and Joos 1922).

[124] Einstein to Kuratorium, 30 April 1920, AMPG, I. Abt., Rep. 1 A, Nr. 1657, p. 125; Einstein to Koch, 5 June 1920, AMPG, I. Abt., Rep. 34, Nr. 7, folder Koch.

[125] See s.v. "Koch, Peter Paul" in Weinmeister (1925–1926), Stobbe (1936–1940), Zaunick and Salié (1956–1962).

[126] Einstein to Kuratorium, 30 April 1920, AMPG, I. Abt., Rep. 1 A, Nr. 1657, p. 125; Franck to Einstein, 4 November 1920, AMPG, I. Abt., Rep. 34, Nr. 2, folder Franck.

[127] Franck to Einstein, 17 April 1920, AMPG, I. Abt., Rep. 34, Nr. 2, folder Franck.

"automatic registration of voltage curves," which Knipping himself had invented.[128] Despite the hint of a possible extension of the research to the impact of electrons on solids, the equipment was mainly used in the investigation of the collisional excitation of atoms in gases.[129] It is worth stressing that in both cases Franck submitted his application "in agreement with" Haber and that the grants for Franck supported research undertaken at Haber's KWI für physikalische Chemie und Elektrochemie which, at the time, was in serious financial trouble, whereas the KWI für Physik in December 1920 still had 77,000 marks available.[130]

Hallwachs, ordinary professor of physics at the Technische Hochschule in Dresden and one of the discoverers of the photoelectric effect, had applied as early as May 1919 for 900 marks to purchase an electrometer to be used by a doctoral candidate to determine "the influence of gas on the long-wave limit of the photoelectric effect."[131] For that purpose he received 1,000 marks in April 1920.[132] Hallwachs's main concern was the systematic collection of data for the purpose of giving as exhaustive a picture as possible[133]; another investigation on the photoelectric effect suggested by Hedwig Kohn, assistant at the Physics Institute of the University of Breslau, addressed new problems of quantum physics. Kohn wanted to perform measurements of the photoelectric effect in gases and vapors in the hope "of testing some consequences of quantum theory and of Bohr's atomic model." For this she needed a high-luminosity quartz spectrograph, which had to be made to order. The application was supported by the institute's director, Otto Lummer, who estimated the costs at between 5,000 and 10,000 marks.[134] In September 1919, Kohn spoke to Einstein about her project and, some months later, sent, together with another supporting letter by Lummer, a more detailed application in which she explained that she wanted to check.

[...] if the photoelectric effect, i.e., ionization, occurs only under irradiation with the light of the ultraviolet limiting line n_∞ of the [spectral] series of the unexcited atoms. This should be the case from the perspective of Bohr's atomic model according to the quantal equation $e \cdot V = h \cdot n_\infty$.[135]

---

[128] Franck to Einstein, 4 November 1920, AMPG, I. Abt., Rep. 34, Nr. 2, folder Franck. On the methods, see Franck and Knipping (1919, 1920), Knipping (1923). On Knipping, see Stintzing (1938).

[129] See e.g., Franck (1921), Franck and Grotrian (1921), Knipping (1921).

[130] Schmidt-Ott to Kuratorium members, 14 December 1920, AMPG, I. Abt., Rep. 1 A, Nr. 1657, p. 152. The KWI für physikalische Chemie und Elektrochemie received 35,000 marks per year both from the Prussian State and the Koppel-Stiftung. In April 1920 it had a deficit of 146,279 marks (vom Brocke 1990, 231, 233).

[131] Hallwachs to Einstein, 2 May 1919, AMPG, I. Abt., Rep. 34, Nr. 4, folder Hallwachs.

[132] Einstein to Kuratorium, 30 April 1920, AMPG, I. Abt., Rep. 1 A, Nr. 1657, p. 125. Actually, with this money Hallwachs purchased, with Einstein's approval, mercury for a McLeod gauge (Hallwachs to Einstein, 13 July 1920, and Einstein to Hallwachs, 19 July 1920, AMPG, I. Abt., Rep. 34, Nr. 4, folder Hallwachs).

[133] For Hallwachs's and his student's research, see Suhrmann (1922).

[134] Kohn to Einstein, with addition by Lummer, 2 August 1919, in Einstein (2004, 124–125, doc. 83).

[135] Kohn to Einstein, 2 January 1920, in Einstein (2004, 337–338, doc. 241); see also Einstein to Kohn, 23 August 1919, AMPG, I. Abt., Rep. 34, Nr. 7, folder Kohn.

Lummer, a well-known expert in optics and photometry who had made important contributions to early research on black-body radiation, also wished to use the spectrograph to extend his investigations using the interference spectroscope in the ultraviolet region, in order to study the "Zeeman phenomenon" and the "dispersion of [light emitted by] gases."[136] At the meeting of 22 April 1920, the Direktorium granted Kohn 7,000 marks. Later on, since the price of the apparatus would probably be higher than previously estimated, the KWI promised to pay any supplement necessary up to a total of 10,000 marks.[137]

In a separate letter Lummer also suggested relaunching research on acoustics, which he considered to be completely neglected in Germany. He estimated that 40–50,000 marks would be needed to equip a well functioning acoustics laboratory, but—aware of the impossibility of receiving such a large sum—asked for only 3,000 marks on behalf of the department of his colleague at the University of Breslau, Erich Waetzmann.[138] Although Einstein expressed a polite interest in Lummer's "valuable suggestions,"[139] they were not taken into consideration at the time. Only in the period 1924/25 was Waetzmann's research on acoustics funded.[140]

The only new area of research receiving money in this period was photochemistry. Fritz Weigert, extraordinary professor of photochemistry at the University of Leipzig, had studied the interaction of linearly polarized light with light-sensitive layers of dye colloids and discovered a new effect: the originally isotropic layers, after illumination, showed permanent properties such as double refraction and dichroism (Weigert 1919a, b, c). In January 1920 he apparently wrote or personally spoke to Einstein about his first observations and, shortly thereafter, applied to the KWI für Physik: "For further quantitative research into the new effects, a very weak double refractiveness must be measured precisely." For this he needed strong mercury light sources and large polarization prisms.[141] Although the subject of photochemistry is closely connected with quantum physics, at this point Weigert did not speak of quantum effects. His observations occurred in what he called the "micellar region," that is, a region of length scales between the molecular dimensions and the wavelength of light, which "is somewhat outside the main area of interest in present physics: the dimensions of atom and nucleus."[142] In fact, the interaction of light with clusters of molecules seemed to be important. Following its policy of diffuse support, the

---

[136]Lummer to Einstein, 2 January 1920, ibid.; (Mehra and Rechenberg 1982, 39–40, vol. 1).

[137]Einstein to Kuratorium, 30 April and 7 December 1920, AMPG, I. Abt., Rep. 1 A, Nr. 1657, pp. 125, 151. See also Kohn to Kuratorium, 2 June 1920, and Kohn to Einstein, 11 July 1920, AMPG, I. Abt., Rep. 34, Nr. 7, folder Kohn.

[138]Lummer to Einstein, 4 August 1919, AMPG, I. Abt., Rep. 34, Nr. 8, folder Lummer.

[139]Einstein to Lummer, 23 August 1919, ibid.

[140]Minutes of the Direktorium meetings, 29 May and 3 July 1924, AMPG, I. Abt., Rep. 1 A, Nr. 1660, pp. 113 and 131 respectively.

[141]Weigert to Einstein, 15 February 1920, AMPG, I. Abt., Rep. 34, Nr. 11, folder Weigert.

[142]Weigert to Einstein, 24 May 1920, AEA, 45–215.

Direktorium granted Weigert 5,000 marks "for his photochemical research."[143] In the following months Weigert published further articles on the new effect.[144] An incidental comment about his own research, "We do not really deal with photochemical processes here," (Weigert 1921, p. 410) is revealing when compared to the Direktorium's motivation. Later, Weigert continued his research on polarized fluorescence in dye solutions and acknowledged the support of the KWI für Physik (Weigert and Käppler 1924).

As something of a curiosity, Rubens was granted an additional 1,000 marks because the money he had previously received for the verification of Planck's law had been stolen.[145]

Rejected Proposals

Again, four applications were rejected because they concerned areas at the fringes of physics. But the opinions among the Direktorium's members about what should be considered "pure physics" were not unanimous. If personal relationships were also a factor, Einstein's favor was clearly not decisive, as highlighted in the case of Seemann, an independent scholar already supported in the previous financial year. Seemann had moved to the X-ray department of the medical clinic at the University of Freiburg and was now working on the "quantitative refinement of the X-ray spectral analysis for therapeutical dosimetry." He needed salt crystals of the highest purity and a Gaede pump.[146] As in May 1919, also on this occasion Einstein was inclined to support Seemann, but the Direktorium voted down his "favorable opinion" and rejected Seemann's application because its purpose was not a "purely physical" one.[147]

In order to enable comparisons with later Direktorium decisions, it is useful to stress the case of Christian Jensen, professor at the Physikalisches Staatslaboratorium of Hamburg, who planned extensive research on polarization of skylight, its causes, its connection to atmospheric electricity, the influence of solar activity, and so on.[148] Jensen had made his request in May 1919, immediately after the first large distribution of money, and received a negative reply due to lack of funds.[149] He then insisted, but his application was definitively rejected a year later "because the KW Institute for Physics must limit itself to the support of investigations of purely physical relevance."[150] In the other cases the rejection was apparently not controversial: Friedrich Schuh, assistant at the Mineralogical-Geological Institute of the University of Rostock, had asked for a scholarship and an Eötvös torsion balance for gravity and

---

[143] Einstein to Kuratorium, 30 April 1920, AMPG, I. Abt., Rep. 1 A, Nr. 1657, p. 125.

[144] See s.v. "Weigert, Fritz" in Weinmeister (1925–1926, 1345–1346, vol. 2).

[145] Einstein to Kuratorium, 7 December 1920, AMPG, I. Abt., Rep. 1 A, Nr. 1657, p. 151.

[146] Seemann to Einstein, 12 July and 28 November 1920, AMPG, I. Abt., Rep. 34, Nr. 10, folder Seemann. Quotation from the latter.

[147] Einstein to Seemann, 10 December 1920, ibid.

[148] Jensen to Einstein, 14 May 1919, AMPG, I. Abt., Rep. 34, Nr. 5, folder Jensen.

[149] Einstein to Jensen, 16 May 1919, ibid.

[150] Jensen to Einstein, 10 June 1919, and Ilse Einstein to Jensen, 7 May 1920, ibid. Quotation from the latter.

magnetic measurements in the field.[151] Siegfried Valentiner, professor of physics at the Bergakademie (School of Mining Engineering) of Clausthal, needed an electric furnace for investigations of the heat conductivity of insulating materials relevant to industry.[152]

Only the project suggested by Gabriele Rabel was rejected for scientific reasons. Rabel had just completed her doctoral dissertation on spectroscopy under Stark's supervision at the University of Greifswald but had previously studied at the University of Berlin, where she had become acquainted with Einstein, and worked on photochemistry (Rabel 1919). In order to explain a phenomenon discovered in her previous research, she planned an experiment with canal rays, that is, beams of positive hydrogen ions deflected by electric and magnetic fields in a vacuum. Usually, Rabel argued, the light seen in such beams resulted from the interaction of the ions with the atoms of the remaining gas in the tube, and not from the interaction of the ions among themselves. With the experimental setup explained in a sketc.h enclosed with the letter, she wanted to establish if light could be emitted by the collision of the hydrogen ions alone. Stark was skeptical about her project but would not hinder her in doing the experiment. According to Rabel he even "generously refrains from requesting funds from the [Kaiser-Wilhelm-] Gesellschaft in favor of those researchers who are not institute directors and are therefore in an even worse situation than he himself."[153] Einstein reacted negatively because a calculation convinced him that the mutual collisions of the hydrogen ions were not abundant enough to lead to an observable emission of light. "To me it appears almost impossible to obtain, by the method you sketc.hed, such a concentration of canal rays that a generation of light by mutual collisions could be observed."[154]

Among the applications left unanswered was one from Born, who had just become director of the Institute for Theoretical Physics at the University of Frankfurt am Main. In his new position he wanted to start a broad program, including measurements of metal vapor pressures with the Knudsen method and experimental investigations of isotopes of chemical elements with the diffusion method, of molecular speed, and of Maxwell's velocity distribution. Also "Tammann's theory of mixed crystals" and "Haber's theory of metals," that is, theories on the structure of solids were to be tested by X-rays. Considering the very limited equipment at his disposal, Born asked for an exceptionally "large sum" without specifying the amount.[155] Two months later Einstein privately let his favorable intentions be known, although he delayed the allocation, waiting for a better time: "I shall happily try to wangle K. W. Institute's funds for your husband when it is possible—when we again have something to

---

[151]Schuh to Einstein, 17 October 1919, Einstein to Schuh, 6 November 1919, and Ilse Einstein to Schuh, 7 May 1920, all in AMPG, I. Abt., Rep. 34, Nr. 10, folder Schuh.

[152]Valentiner to Einstein, 16 May 1919, and Einstein to Valentiner, 20 May 1919, AMPG, I. Abt., Rep. 34, Nr. 11, folder Valentiner.

[153]Rabel to Einstein, 3 November 1919, AMPG, I. Abt., Rep. 34, Nr. 9, folder Rabel.

[154]Einstein to Rabel, 20 November 1919, ibid.

[155]Born to Einstein, 1 July 1919, AMPG, I. Abt., Rep. 34, Nr. 1, folder Born.

distribute."[156] In fact, as already reported, at the end of September 1919 the KWI still had roughly 65,000 marks to spend, but the Direktorium had probably decided not to supply the entire laboratory equipment of a university institute, after the criticism it had received from Siemens and Krüss regarding the previous distribution. Born must have found another way to get the money, since Ilse Einstein later noted on Born's application "not needed any longer" and the request was not taken into consideration at the Direktorium meeting of 22 April 1920.

Finally, a "secret" project "the aim of which is to gain electric energy directly from heat" was proposed by a supporting member of the KWG.[157] Einstein was willing to referee the project because he thought "it improbable but not absolutely impossible that a useable idea is presented here for a solution of the problems raised in the letter."[158] The idea must have been discarded, though, for there is no further correspondence relating to it.[159]

Annual Report 1920/21

The report concerning the period from 1 April 1920 to 31 March 1921 is remarkable for its lack of commitment:

> The amount of 59,800 marks distributed during the budget year 1920/21 is relatively small. We preferred to increase in this way the stock of the available funds, in order to be able eventually to cope with some single larger and more expensive tasks. The sum spent was used to enable a greater number of important physical investigations. In most of the cases, the money was used for buying equipment which remains the property of the Kaiser Wilhelm Institute.[160]

The declared intention was still the support of large-scale scientific research, but since such research, involving more than one institution, had neither been proposed from the outside nor by the Direktorium itself, the original plan became only a hope for the future.

This time nobody in the KWI für Physik made any trouble. Siemens had died in October 1919. In the following years all Direktorium decisions continued to be approved without objection. If we exclude Freundlich, none of the projects supported in 1920/21 had been directly suggested by Einstein, who only showed particular interest in the work of Hettner and the measurements of Regener and Baeyer. Only in two cases, those of Hettner and Franck, can we establish a direct relation between an application and a member of the Direktorium. As mentioned, Hettner was an assistant to Rubens and had been working with him for many years. Franck was

---

[156]Einstein to Hedwig Born, 1 September 1919, in Einstein and Born (1969, 30).

[157]Josef Kaiser to KWG, 5 March 1920, and von Harnack to Einstein, 31 March 1920, AMPG, I. Abt., Rep. 34, Nr. 6, folder Kaiser.

[158]Einstein to von Harnack, 5 April 1920, ibid.

[159]The other unanswered requests were those of Ehrenhaft (see page 49) and of Lummer concerning acoustics (see page 60).

[160]Report of the KWI für Physik 1 April 1920–31 March 1921, AMPG, I. Abt., Rep. 34, Nr. 12. The report was sent to the KWG on 13 July 1921 (see Einstein to von Harnack, 13 July 1921, AMPG, I. Abt., Rep. 34, Nr. 13).

a collaborator of Haber and had discussed his project with him before submission. However, this does not prove that the research had been suggested by Rubens or Haber themselves. All the other projects were clearly proposed by university researchers on their own initiative. The institute thus confirmed its role as a grant distributor without priorities.

Freundlich, after three years on the payroll of the KWI für Physik, had finally obtained a permanent position. Einstein, in the annual report referred to above, laconically stated:

> The only scientific member of the institute until now, Dr. Freundlich, continued the work mentioned in the last report; he left the Kaiser Wilhelm Institute at the end of the calendar year 1920 and continued his work in the position of an observer at the Astrophysikalisches Observatorium in Potsdam.[161]

In fact, an agreement had been reached between the Observatory's director, Gustav Müller, and the Prussian Ministry of Education according to which "Freundlich, until further notice, will be charged primarily with the work on Einstein's relativity theory and will be employed for other tasks only to the extent that these will not hinder him in his main duty."[162]

During the fiscal year 1920/21 the institute had spent little more than in the previous one,[163] that is a total of 89,476 marks, of which 71,600 were for "scientific purposes" and 4,500 marks for Freundlich's salary for the period April–December 1920.[164] In March 1921 the KWG doubled its contribution from 50,000 to 100,000 marks, retroactively for the year 1920/21 and for the following one.[165] In June 1920 Einstein's annual salary had already been doubled from 5,000 to 10,000 marks as of 1 April 1920.[166] Taking into account the latest transfers by the KWG, but not the grants approved on 3 March 1921, at the end of March 1921 the institute's account showed a balance of 138,794 marks.[167] Meanwhile, inflation had started to erode the value of the capital bestowed to the KWG. Since April 1920, the revenue coming from interest on the capital and membership fees was not enough to cover the running expenses of the KW Institutes. Increasing allocations by German big business, by the newly founded Notgemeinschaft der Deutschen Wissenschaft, and mainly by the national and the Prussian governments, helped to ease the science funding situation

---

[161] Report of the KWI für Physik 1 April 1920–31 March 1921, AMPG, I. Abt., Rep. 34, Nr. 12.

[162] Prussian Ministry of Education to Müller, 1 June 1920, GStA, I. HA, Rep. 76 76 V c, Sekt. 1, Tit. 11, Teil 2, Nr. 6 b, Bd. 8, p. 165.

[163] For comparison see Tables 3.1 and 3.2.

[164] Survey of income and expenditure 1 April 1920–31 March 1921, AMPG, I. Abt., Rep. 34, Nr. 13. The survey, dated 9 June 1921, was sent to the KWG on 13 July 1921 (see Einstein to von Harnack, 13 July 1921, ibid.). We cannot explain the discrepancy between the report and the survey concerning expenditures for scientific purposes.

[165] Von Harnack to Einstein, 16 March 1921, AMPG, I. Abt., Rep. 34, Nr. 13.

[166] Schmidt-Ott to Einstein, 23 June 1920, ibid.

[167] Bank statement 1 January–31 March 1921, AMPG, I. Abt., Rep. 34, Nr. 8, folder Mendelssohn.

in 1921.[168] While almost all the research institutes showed increasing deficits, the KWI für Physik was accumulating interest.

The Period 1921/22. Projects Funded

In the following budgetary period, April 1921–March 1922, the cautious fund distribution continued. The Direktorium held four meetings on 3 March, 7 July, 12 October 1921, and 26 January 1922.[169] Only an application from an unknown inventor was rejected; all other requests received were granted, a total of twelve grants.

Some allocations concerned holdovers from the previous year. Fürth at last received the 2,000 marks he had requested for his measurements of the elementary electric charge. Rubens's "radiation measurements" were financed with an additional 1,000 marks. The journal *Physikalische Berichte* again received 5,000 marks, but "for the last time."[170] A further 850 marks from the grant approved in 1919/20 were paid out to Steubing, who also made a new request: he asked for a high-voltage rectifier for a new investigation of the Stark effect in band spectra and subsequently received 10,000 marks.[171] Wagner's new request for support to continue his investigation of the "influence of potential and anticathode material on the emission of X-rays"[172] was also granted an additional 2,500 marks.[173]

Försterling, "because of the circumstances" in Danzig, had not yet been able to perform his research on the temperature dependence of constants in metals, for which he had received a grant in 1919. As he was about to move to Jena, where it would not be possible to carry out the original project, he asked whether he could spend the money for a different investigation and proposed two alternatives: either he might study the Paschen-Back effect in hydrogen spectra or the dilatation of $NaCl$ and similar crystals at very low temperatures.[174] The first topic would have complemented spectroscopic research supported in other cases: the Paschen-Back effect consists in the merging, caused by a strong magnetic field, of the anomalous Zeeman splitting of spectral lines into a normal splitting (normal Zeeman effect). The Direktorium granted Försterling "total freedom" in the use of the money, thus proving that it set no priorities itself.[175] Försterling then worked on the Zeeman/Paschen-Back effect

---

[168] Witt (1990), vom Brocke (1990, 227–238). On the Notgemeinschaft der Deutschen Wissenschaft, see Marsch (1994), Richter (1972).

[169] Einstein to Kuratorium, 7 March, 20 October 1921, and 28 January 1922, AMPG, I. Abt., Rep. 1 A, Nr. 1658, pp. 166–167, 181, 188 [bis] respectively; minutes of the Direktorium meeting, 26 January 1922, AMPG, I. Abt., Rep. 34, Nr. 12; KWI für Physik to Försterling, 16 July 1921, AMPG, I. Abt., Rep. 34, Nr. 2, folder Försterling; (Kirsten and Treder 1979, 153–154, vol. 1).

[170] For all these allocations, see Einstein to Kuratorium, 7 March 1921, AMPG, I. Abt., Rep. 1 A, Nr. 1658, pp. 166–167.

[171] Einstein to Kuratorium, 28 January 1922, AMPG, I. Abt., Rep. 1 A, Nr. 1658, p. 188 [bis].

[172] Wagner to Einstein, 22 October 1921, AMPG, I. Abt., Rep. 34, Nr. 11, folder Wagner. Wagner refers to the work exposed in Kulenkampff (1922).

[173] Minutes of the Direktorium meeting, 26 January 1922, AMPG, I. Abt., Rep. 34, Nr. 12.

[174] Försterling to Einstein, 3 April 1921, AMPG, I. Abt., Rep. 34, Nr. 2, folder Försterling.

[175] KWI für Physik to Försterling, 16 July 1921, ibid.

in hydrogen and published his results two years later in a paper in which he thanked the KWI für Physik (Försterling and Hansen 1923).

Another new project in spectroscopy, which turned out to have important consequences for quantum physics, was supported with 10,000 marks.[176] The request for support for an "investigation on the band spectra of mono-atomic metal vapors"[177]— that was not further explained—was submitted by Walther Gerlach, an expert on thermal radiation, who had just been appointed as extraordinary professor of experimental physics at the University of Frankfurt am Main (Hoyer 1993). He had been advised to apply to the KWI für Physik by Franck and by Born, who was leaving Frankfurt for Göttingen. In the summer of 1921, Gerlach and Otto Stern, a former collaborator of Einstein in Prague and Zurich and now extraordinary professor of theoretical physics in Frankfurt, were using such a metal vapor, specifically one consisting of silver atoms, to check a prediction made by Stern. By deflecting beams of silver atoms with magnetic fields they wanted to test whether the magnetic moment generated by an orbiting electron in an atom led to directional quantization.[178] The tests were carried out mainly by Gerlach until spring of 1922, as Stern had taken over a chair in Rostock. The results of this well-known "Stern-Gerlach experiment" confirming the quantum-theoretical prediction were published in three consecutive papers and a recapitulating report, in which the authors thanked the KWI für Physik, and especially its Director Einstein, for the support received, as well as other donors (Gerlach and Stern 1922a, b, c, 1924).

Another investigation that would allegedly be of "utmost importance for the foundations of quantum theory" was proposed by Max Trautz, extraordinary professor of physical chemistry at the University of Heidelberg.[179] Trautz planned a series of measurements of the specific heat of many gases and for large temperature ranges using a new, very precise, and quickly working method of his own invention. To make it very clear who was and who might be interested, he added to his letter: "Nernst, to whom, next to Einstein and Planck, we probably owe the greatest advancement of research on specific heats in the past decades, is informed in detail about our investigations and aims."[180] However, name dropping was not enough to obtain approval. Trautz was asked for more detailed information about his method and in the end received only half of the 6,000 marks he had requested.[181] A short handwritten note by Einstein on Trautz's first letter, "More precise information about the method. Evaluation

---

[176]Einstein to Kuratorium, 20 October 1921, AMPG, I. Abt., Rep. 1 A, Nr. 1658, p. 181.

[177]Gerlach to Einstein, 27 July 1921, AMPG, I. Abt., Rep. 34, Nr. 3, folder Gerlach.

[178]For the history of the experiment, see Mehra and Rechenberg (1982, 433–443, pt. 2, vol. 1).

[179]Trautz to Einstein, 25 March 1921, AMPG, I. Abt., Rep. 34, Nr. 11, folder Trautz.

[180]Ibid.

[181]KWI für Physik to Trautz, 16 July 1921; Trautz to Einstein, 5 December1921; Einstein to Trautz, 30 January 1922, all in AMPG, I. Abt., Rep. 34, Nr. 11, folder Trautz; minutes of the Direktorium meeting, 26 January 1922, AMPG, I. Abt., Rep. 34, Nr. 12.

to me,"[182] gives an insight into the decision procedures of the Direktorium. Trautz's systematic collection of data lasted for several years and was supported by the KWI für Physik again in 1923.[183]

As to research on the properties of specific material systems, an investigation initiated by Clemens Schaefer, ordinary professor of experimental physics at the University of Marburg, was supported with 10,000 marks.[184] Schaefer, a former student of Warburg and assistant of Rubens, was an esteemed theoretical as well as experimental physicist. He had been working on the infrared spectroscopy of crystals for a couple of years (Bergmann 1958). In particular, his ongoing investigation focused on the infrared eigenvibrations of crystalline silicates, in order to draw conclusions concerning the structure of the material.[185] At first, Schaefer had applied for money to Planck as the Secretary of the Prussian Academy of Sciences, who in turn suggested that he apply to Einstein and the KWI für Physik. There was some bitter haggling with Schaefer, who felt unduly constrained by the conditions set by the Direktorium as to the use of the funds. While Schaefer wished to use them for running costs and for buying crystals, the KWI insisted that they could be used only for buying equipment, of which it would retain ownership, and promised more money only if needed.[186] Perhaps this was the reason why Schaefer did not acknowledge the institute's support in his paper on silicates (Schaefer and Schubert 1922).

In view of the KWI's inflexibility with Schaefer, it is surprising how smoothly the grant for Albert Wigand was approved. Wigand, extraordinary professor of atmospheric physics at the University of Halle, had asked for money to cover the running costs "for the continuation of my physical investigations during air flights."[187] Nernst, Wilhelm Wien, and Gustav Mie had advised him to approach the KWI für Physik. Wigand was the leader of a team of researchers working for many years on atmospheric physics. During flights with balloons and airplanes, they performed systematic measurements of a wide range of phenomena, focusing in the last years on atmospheric electricity and radioactivity in particular (Hergert 1993; Kolhörster 1933; Wigand 1924). At that time Nernst was taking an increasing interest in atmospheric electricity as well as in cosmic ray physics. Not accepting the idea of the unavoidable "heat death" (Nernst 1921a, 1) of the universe, he was looking for any phenomenon suggesting the existence of a balance between energy degradation and energy creation in the cosmos.[188] We may assume that this was the reason why

---

[182] Written on the letter of Trautz to Einstein, 25 March 1921, AMPG, I. Abt., Rep. 34, Nr. 11, folder Trautz.

[183] Laue to Kuratorium, 6 July 1923, AMPG, I. Abt. Rep. 1 A, Nr. 1659, p. 41a. The method and some first results are exposed in Trautz and Grosskinsky (1922).

[184] Einstein to Kuratorium, 7 March 1921, AMPG, I. Abt., Rep. 1 A, Nr. 1658, pp. 166–167.

[185] Schaefer to Einstein, 9 January 1921, AMPG, I. Abt., Rep. 34, Nr. 10, folder Schaefer.

[186] Schaefer to Schmidt-Ott, 13 May 1921; Schmidt-Ott to Planck, with additions by Planck and Einstein, 25–29 May 1921; KWI für Physik to Schaefer, 16 July 1921; and Schaefer to Einstein, 11 August 1921, all in AMPG, I. Abt., Rep. 34, Nr. 10, folder Schaefer; Planck to Schaefer, 14 June 1921, AMPG, I. Abt., Rep. 1 A, Nr. 1658, p. 180.

[187] Wigand to Einstein, 9 January 1922, AMPG I. Abt., Rep. 34, Nr. 11, folder Wigand.

[188] See also Bartel and Huebener (2007, 306–326), Gunther (1924a), Hiebert (1978, 448–449).

Wigand received 2,000 marks for research that could have easily been considered not of "purely physical relevance," as in the case of Jensen's investigation of skylight polarization.

Einstein, too, used institute funds, albeit moderately, for his own scientific interests. His collaborator Grommer was paid 2,000 marks for "theoretical research in the field of relativity theory."[189] It is not possible to establish a direct connection between this "research" and Einstein and Grommer's paper on Theodor Kaluza's unified field theory published in 1923 (Einstein and Grommer 1923); nevertheless, it seems plausible that Grommer's help was needed for solving mathematical problems arising from the extension of general relativity towards a unified field theory, toward which Einstein was working at this time.[190] In January 1922, Einstein granted to himself the modest sum of 579.85 marks "for a mercury condensation pump."[191] In all likelihood, the pump had been used for an experiment suggested by Einstein and carried out in December 1921 by Geiger and Walther Bothe in the laboratory of the Physikalisch-Technische Reichsanstalt.[192] Since the summer of 1921, Einstein had been pondering an experiment that he hoped would disprove the wavelike nature of the radiation field.[193] The reasoning went like this: According to corpuscular theory, the frequency of emitted light should not depend upon the state of motion of the emitting particle. According to the wave theory, though, the frequency, and therefore the color of light emitted by a moving particle, depends on the direction of the emission because of the Doppler effect. Einstein proposed to check whether the light emitted by canal rays shows a continuous spectrum of colors or is monochromatic (Einstein 1921). It turned out that, in Einstein's words and interpretation,

> [...] the light emitted by the moving canal ray particle is strictly monochromatic while, according to the wave theory, the color of the elementary emission should be different in the different directions. By that it is positively proved that the undulatory field has no real existence and that the Bohr emission is an instantaneous process in the proper sense.[194]

However, Laue and Ehrenfest objected strongly to this interpretation by pointing to the fact, among others, that, according to the wave theory, light is emitted in short pulses which would explain why a color shift is not observable. After some hesitation, Einstein finally admitted that it was not possible to draw from the experiment "consequences about the nature of the elementary emission process" (Einstein 1922c, 22).

---

[189]Einstein to Kuratorium, 7 March 1921, AMPG, I. Abt., Rep. 1 A, Nr. 1658, pp. 166–167.

[190]Pais (1982, 328–329). For an exposition of Einstein's work towards a unified field theory in the scientific context of the time, see Goenner (2004).

[191]Minutes of the Direktorium meeting, 26 January 1922, AMPG, I. Abt., Rep. 34, Nr. 12; Einstein to Kuratorium, 28 January 1922, AMPG, I. Abt., Rep. 1 A, Nr. 1658, p. 188 [bis].

[192]See the bill for the pump addressed to Geiger on 28 December 1921, AMPG, I. Abt., Rep. 34, Nr. 3, folder Geiger.

[193]See, e.g., Einstein to Sommerfeld, 27 September 1921, in Einstein and Sommerfeld (1968, 91). For a description of the experiment and its interpretation, see Mehra and Rechenberg (1982, 516–518, vol. 1).

[194]Einstein to Born, 30 December 1921, in Einstein and Born (1969, 95–96).

Finally, in this period, a very important decision regarding the distribution policy was made that would definitively change the institute's rationale. Franck, who had become ordinary professor for experimental physics in Göttingen, had to leave most of his equipment in Berlin. Therefore, he asked for 20,000 marks in order to "continue research on the collision of electrons and fluorescence."[195] The reason for the large sum was that he wanted to perform a parallel investigation by optical means, but the money granted by the Prussian Ministry of Education for the modernization of the University's equipment was insufficient. Einstein received Franck's request a few days after the first Direktorium meeting of the new fiscal year, in which 20,000 marks had been distributed. But the institute still had more than 60,000 marks at its disposal[196] and at the end of the month, as mentioned above, the budget was doubled. Nevertheless, in what seems a recurrence of Grebe and Bachem's case of December 1919 and probably because of Krüss's criticism concerning the distribution policy from April 1919, Einstein took pains to ask the industrialist Franz Oppenheim to provide not only 20,000 marks to Franck but also 10,000 marks to the engineer H. Boas of Berlin, for the construction of "a new kind of spectral heliograph important for physical astronomy."[197] At the same time, in spring 1921, Born also joined Franck and Pohl in Göttingen as ordinary professor for theoretical physics.[198] Soon after he, too, applied to the KWI für Physik for 25,000 marks in order to buy X-ray instruments.[199] At first, the Direktorium asked for more details on the planned research and for a list of the apparatuses needed.[200] Einstein, although inclined to support Born, asked for patience because the requested sum "would gobble" the greater part of the budget.[201] Some months later, in October, Born again asked the "powerful Director" of the KWI für Physik, also in Franck's and Pohl's name, for a large sum of money. A quick decision had become necessary because the instruments had already been ordered in the hope that the KWI would agree.[202] This time, the objections that still hindered a decision in Franck's favor in March must have been overcome and the grant was approved, as we may conclude from Born's letter of thanks.[203] The Direktorium allocated 100,000 marks.[204] The grant was not connected

---

[195]Franck to Einstein, 9 March 1921, AEA, 40–137.

[196]Einstein to Kuratorium, 7 March 1921, AMPG, I. Abt., Rep. 1 A, Nr. 1658, pp. 166–167; Schmidt-Ott to Kuratorium members, 16 March 1921, ibid., p. 168.

[197]Einstein to Oppenheim, 15 March 1921, in Einstein (2009, 143, doc. 99). We do not have more information about Boas and his project.

[198]On their work in Göttingen, see Hund et al. (1988).

[199]KWG to Einstein, 6 June 1921, AMPG, I. Abt., Rep. 1 A, Nr. 1658, p. 178. Born's application is missing.

[200]Ilse Einstein to Born, 16 July 1921, AMPG, I. Abt., Rep. 34, Nr. 1, folder Born.

[201]Einstein to Born, 22 August 1921, in Einstein and Born (1969, 85).

[202]Born to Einstein, 21 October 1921, ibid., pp. 86–88. See also Born's comments on the difficult economical situation, ibid., pp. 88–89.

[203]Born to Einstein, 29 November 1921, ibid., pp. 91–93. There is no document attesting to a formal decision.

[204]"Verzeichnis der im Rechnungsjahre 1921/1922 bewilligten Zuwendungen," AMPG, I. Abt., Rep. 34, Nr. 13; Laue to Schmidt-Ott, 15 January 1923, AMPG, I. Abt., Rep. 1 A, Nr. 1659, p. 20.

to a specified research project but, in evident breach of the declared funding policy, was given to a university institute for whatever its research activities might require.[205]

Two applicants were asked for further details.[206] The correspondence relating to Seeliger's new application is not available. His request was approved in the following fiscal year. Koenigsberger, extraordinary professor of mathematical physics at the University of Freiburg, had preliminarily asked Einstein whether his application would stand a chance. He wanted to continue his research "also covering to some extent geophysics,"[207] which was his main area of interest. As he claimed, he was boycotted, probably "because of my political stand," by Lenard, who decided about the appropriations of the Academy of Science of Heidelberg. Furthermore, Koenigsberger had to "fight against many difficulties" in his institute. He was one of these professors who suffered under the undivided power of the director of the institute to which he belonged (Jungnickel and McCormmach 1986, 287–289, vol. 2). The hint at the political background of his difficulties was not casual: Koenigsberger was member of the local parliament for the Social Democratic Party (Schröder 1995, 559) and knew that Einstein was a political friend. Both he and Einstein had been members of the pacifist organization Bund Neues Vaterland (New Fatherland League) during the First World War.[208] In addition, ten years earlier Einstein had ranked Koenigsberger as one of the best young physicists in Germany (Einstein 1993b, 308, doc. 275). For all of these reasons, perhaps, Einstein did not give an immediate negative answer but instead suggested to Koenigsberger that he submit an application with "a more precise characterization of your planned research (putting the stress on the mere physical aspects)."[209] His case was approved in the following fiscal year.

The Rejected Proposal

The only application rejected came from an outsider. In February 1921, Eduard Schweigler from Vienna, probably an engineer, submitted a project for the construction of a "Sehzelle" (vision cell), an apparatus for the production and transmission of light signals using the photoelectric effect.[210] Within a week Einstein replied that the KWI für Physik could "support only purely scientific aims" and added:

> Privately, I must observe that I have not understood the purpose and functioning of your "organ for vision." Normal light does not remove electrons from a silver mirror. Even if particular surfaces sensitive to light were to be used, it would be very difficult to reach sufficient sensitivity.[211]

---

[205] In view of this substantial support, it is somewhat surprising that of twenty-five publications by Franck's group between 1922 and 1924, none contains an acknowledgment of the KWI für Physik whereas other donors are thanked several times.

[206] Minutes of the Direktorium meeting, 26 January 1922, AMPG, I. Abt., Rep. 34, Nr. 12.

[207] Koenigsberger to Einstein, 2 November 1921, AMPG, I. Abt., Rep. 34, Nr. 7, folder Königsberger.

[208] Invitation to the meeting of Bund Neues Vaterland of 16 August 1915, Bundesarchiv Koblenz, NL 199 Wehberg, Nr. 14, p. 166. On Einstein's pacifist activities in the Bund, see Goenner and Castagnetti (1996).

[209] Einstein to Koenigsberger, 23 January 1922, AMPG, I. Abt., Rep. 34, Nr. 7, folder Königsberger.

[210] Schweigler to Einstein, 11 February 1921, in Einstein (2009, 78–79, doc. 46).

[211] Einstein to Schweigler, 17 February 1921, in Einstein (2009, 87–88, doc. 51).

Annual Report 1921/22

In the report[212] for the period April 1921–March 1922 Einstein, after declaring that the institute had supported research "in the most different fields of physics," singles out three investigations, pertaining to atomic physics and black-body radiation, and situates them in the broader scientific context:

> Whereas Laue's method of X-ray interference leads to the knowledge of the geometrical constitution of solids, the investigation of spectral properties in the infrared, brought forward particularly by Rubens, may give us information about the *dynamical* behavior of atoms. Clemens Schaefer [...] has set out to ascertain in this way the chemical structure of silicates important for the constitution of the Earth's crust.

Thereafter, Einstein mentions Wagner's "outstanding precision work" in the "quantitative investigation of the generation of X-rays by cathode rays," which is "of high importance for atom theory and the fundamental questions of quantum theory." Finally, Einstein explains that "[a]fter a critical investigation by Nernst and Wulf had raised doubts about the strict validity of the fundamentally important Planck's formula, a test as precise as possible seemed imperative." Therefore, the institute financed the "precision measurements by the method of isochromates in the infrared" through which Rubens "has verified Planck's formula with higher precision than hitherto possible." All three undertakings had been funded for more than one budgetary period. As to the other research projects, most of them, despite Einstein's introductory remarks, concerned quantum physics either directly or indirectly. As in the previous years, the projects were not proposed by the Direktorium members but by university researchers and were a clear sign of the shifting interests of the physics community.

In the end, the sum distributed for scientific purposes was very small: about 88,000 marks.[213] On 31 March 1922, the surplus amounted to 197,000 marks. This has to be compared to the 306,000 marks at the institute's disposal during that year, including the balance of the previous year. The KWI für Physik had not even been able to support research amounting to the total of its incoming funds, not to mention its reserves.

Shortly before the end of the budget year, the KWG increased its contribution to the research funds of 50,000 marks retroactively as of 1 October 1921, and to 261,000 marks yearly for the budget year 1922/23. Einstein's salary was also raised from 10,000 to 20,000 marks (18,000 marks net) and that of his secretary Ilse from 3,000 to 8,000 marks, both as of 1 October 1921.[214] Including the annual 25,000

---

[212]Report of the KWI für Physik 1 April 1921–31 March 1922, AMPG, I. Abt., Rep. 1 A, Nr. 1665, pp. 57–58. The report was sent to the KWG in May 1922 (see Einstein to von Harnack, 22 May 1922, ibid., p. 56). Underlining by Einstein.

[213]The survey of income and expenditure 1 April 1921–31 March 1922 (AMPG, I. Abt., Rep. 1 A, Nr. 1665, p. 59) gives other figures than those we have calculated but, as a KWG auditor noted on the document, the survey is not correct and its data do not fit those of the bank statements. The survey was sent to the KWG in May 1922 (see Einstein to von Harnack, 22 May 1922, AMPG, I. Abt., Rep. 1 A, Nr. 1665, p. 56).

[214]The increments were decided by the Senat of the KWG on 24 March 1922 (von Harnack to Einstein, 28 March 1922, AMPG, I. Abt., Rep. 34, Nr. 2, folder Einstein).

marks from the Koppel-Stiftung, but not the interests on the money saved from the previous years, for the period 1922/23 the KWI für Physik could therefore reckon with 314,000 marks in new income.

3.4   The Period 1922/23

The budgetary period from April 1922 to March 1923 turned out to be different from the previous ones. Einstein acted as the director of the institute only until July 1922. From then on he traveled until March 1923, and Laue, who had become a member of the Direktorium not long before,[215] took over his duties. At the start of the period, the Direktorium increased the frequency of its meetings and met on 18 May, 15 June, and 13 July 1922, that is, once a month. However, after Einstein's departure further applications were approved by correspondence or in informal meetings at the Academy. Eleven single researchers or groups were supported with grants of varying amounts, ranging from 1,500 to 300,000 marks.[216] This difference in the nominal value is not particularly significant, since inflation was accelerating dramatically.[217] Only one request was not granted during this period.

Projects Funded

Overall business continued as usual. In several cases the appropriations were just supplements to grants issued in previous years. The scientists needed extra funding because of rising inflation. For his work on the Stark effect Steubing received 2,500 marks in addition to the sum granted shortly before.[218] Schaefer also received an additional 7,500 marks for the instrument already approved in March 1921.[219] Pohl, and with him the Physics Institute of Göttingen University, was supported with an

---

[215]Laue was co-opted by the Direktorium on 3 October 1921 (Einstein to Kuratorium, 20 October 1921, AMPG, I. Abt., Rep. 1 A, Nr. 1658, p. 181). His election was approved by the Kuratorium some weeks later (Schmidt-Ott to Einstein, 2 December 1921, AMPG, I. Abt., Rep. 34, Nr. 8, folder Laue).

[216]Minutes of the Direktorium meetings, 18 May and 15 June 1922, AMPG, I. Abt., Rep. 34, Nr. 12; Einstein to Kuratorium, 26 and 28 July 1922, AMPG, I. Abt., Rep. 1 A, Nr. 1658, pp. 203a and 205 respectively; Einstein to Kuratorium, 15 September 1922, AMPG, I. Abt., Rep. 1 A, Nr. 1659, p. 1 bis; Ilse Einstein to Kuratorium, 30 November 1922, ibid., p. 11; Laue to Kuratorium, 15 January and 8 March 1923, ibid., pp. 20 and 26 respectively.

[217]The exchange rate rose from 7.43 marks for 1 US$ at the end of the First World War in November 1918 to 10.39 marks in March 1919, 83.89 marks in March 1920, 62.45 marks in March 1921, and to 284.19 marks in March 1922. In July 1922 the rate was already 493.22 marks, in November 7,183.10 marks, in January 1923 17,972 marks, in March 21,190 marks. The cost-of-living index, set to 1 in 1913, rose from 9.6 in March 1920 to 11.4 in March 1921, 29 in March 1922, 2,854 in March 1923 (Laursen and Pedersen 1964, 133–135). For the consequences of inflation on the financing of scientific research, see Forman (1968, 206–358). As to the impact of inflation on the KWG, see vom Brocke (1990, 198–201).

[218]Minutes of the Direktorium meetings, 18 May 1922, AMPG, I. Abt., Rep. 34, Nr. 12; Einstein to Kuratorium, 12 June 1922, AMPG, I. Abt., Rep. 1 A, Nr. 1658, p. 197.

[219]Einstein to Kuratorium, 28 July 1922, AMPG, I. Abt., Rep. 1 A, Nr. 1658, p. 205.

additional 90,000 marks.[220] Kohn, too, requested more money because the price for the quartz spectrograph had soared to six times the amount approved for its purchase in 1920.[221] The manufacturer had very cleverly inserted into the sales contract a clause which left the price open until delivery.[222] After some debate in which Warburg argued against further support,[223] and after an unsuccessful attempt by Einstein to negotiate a rebate,[224] the Direktorium supplied an additional 43,000 marks.[225] The instrument was finally delivered in August 1923,[226] but the research was still not concluded as late as March 1926.[227] What in 1919 had seemed to be an advanced research project worthy of support, was in 1926 no longer of particular benefit to the new quantum theories of Heisenberg and Schrödinger.

In a few other cases, the applicants were old acquaintances with new projects aiming to continue research programs that had already been approved in previous years. Seeliger was given 1,500 marks for a high-voltage battery for his systematic investigation of glow discharges in gases.[228] Stern, who meanwhile had become extraordinary professor of theoretical physics at the University of Rostock, also asked for 20,000 marks to continue his "experiments concerning the magnetic properties of silver atoms."[229] An application with a detailed outline of the project is missing, but the Direktorium granted the sum for "apparatuses for magnetic and electric research on molecular beams."[230] Actually, during 1922, Stern thought about whether the directional quantization could also be detected as the effect of an inhomogeneous electric field (Stern 1922). The planned experimental work started only after Stern moved to Hamburg in January 1923 and led to fundamental results for quantum physics (Estermann 1976, 42–43).

Another continuation of research previously funded by the KWI für Physik, that is, Franck's experiments involving the collision of electrons with atoms, was suggested by two assistants at Haber's institute, Hartmut Kallmann and Knipping, whom we already know from the research project under Franck's supervision supported in the period 1920/21. They needed a highly sensitive electrometer for detecting the ions formed by the impact of the electrons. Haber recommended the application with a

---

[220]Laue to Kuratorium, 15 January 1923, AMPG, I. Abt., Rep. 1 A, Nr. 1659, p. 20.

[221]Kohn to Einstein, 20 May 1922, AMPG, I. Abt., Rep. 34, Nr. 7, folder Kohn.

[222]Kohn to Haber, 31 May 1922, ibid.

[223]Warburg to Einstein, 2 June 1922, ibid.

[224]Einstein to Schmidt & Haensch, 17 June 1922, ibid.

[225]Minutes of the Direktorium meeting, 15 June 1922, AMPG, I. Abt., Rep. 34, Nr. 12; Einstein to Kuratorium, 26 July 1922, AMPG, I. Abt., Rep. 1 A, Nr. 1658, p. 203a.

[226]Kohn to Laue, 16 August 1923, AMPG, I. Abt., Rep. 34, Nr. 7, folder Kohn.

[227]Kohn to Laue, 20 March 1926, ibid. Other documents concerning this case are in the same folder.

[228]Minutes of the Direktorium meeting, 18 May 1922, AMPG, I. Abt., Rep. 34, Nr. 12; Einstein to Kuratorium, 12 June 1922, AMPG, I. Abt., Rep. 1 A, Nr. 1658, p. 197.

[229]Einstein to the Direktorium members, 14 May 1922, AMPG, I. Abt., Rep. 34, Nr. 12.

[230]Minutes of the Direktorium meeting, 18 May 1922, ibid.

note written on the proposal.[231] Nernst and Warburg objected to the purchase because of the high price of 40,000 marks, while Laue and Einstein were in favor.[232] In the end, three months later, Haber received only half of the requested amount,[233] despite his protest: "In my opinion 100 dollars is not too much for such an instrument. 40,000 marks, today, corresponds to 100 dollars and will likely mean even less in three to four months."[234] He ordered the instrument on the same day and sent a copy of the order to the KWI für Physik.

The new research planned by Koenigsberger, too, fit with the other investigations on the atomic structure funded in the preceding years. As reported, Koenigsberger had been advised by Einstein to stress in his application "merely the physical aspects" of his intended research. Therefore, instead of working on geophysics he decided to resume experiments with beams of positive hydrogen ions (canal rays) scattered by gas atoms. Such investigations, he thought, "could perhaps provide information on the distribution of the electrons around the nuclei of simple atoms and supplement the conclusions drawn from X-ray interference images in lithium."[235] With this motivation, Koenigsberger was immediately granted the requested 4,000 marks to buy a transformer.[236] In fact, the experiments, which brought new insights into the absorption of canal rays by gases, were carried out some years later by Koenigsberger's student Conrad (Conrad 1926a; Conrad 1926b).

Although there is no written application giving more information, we can reasonably suppose that the research for which Pringsheim received the institute's support concerned topics relating to the quantum theory of light. The KWI granted him 6,530 marks to buy quartz absorption cells for "his work on light absorption in mercury vapor."[237] Pringsheim, a former student of Röntgen, was extraordinary professor of physics at the University of Berlin and a highly esteemed member of the Berlin physics community. In the preceding years, together with Pohl and Franck, he had eminently contributed to deepening the knowledge on luminescence on the basis of atomic and quantum theory.[238] In 1922 in particular, he was working on fluorescence phenomena whose quantum theoretical interpretation was a subject of controversy

---

[231] Kallmann and Knipping to KWI für Physik, with addition by Haber, 3 June 1922, AMPG, I. Abt., Rep. 34, Nr. 6, folder Kallmann.

[232] Einstein to the Direktorium members, with addition by Nernst, 1–4 July 1922; Laue to KWI für Physik, 3 July 1922; Warburg to Einstein, 7 July 1922, all in AMPG, I. Abt., Rep. 34, Nr. 13; Einstein to Laue, 12 July 1922, AMPG, I. Abt., Rep. 34, Nr. 2, folder Einstein.

[233] Einstein to Kuratorium, 15 September 1922, AMPG, I. Abt., Rep. 1 A, Nr. 1659, p. 1 bis.

[234] Haber to KWI für Physik, 6 July 1922, AMPG, I. Abt., Rep. 34, Nr. 13. On Kallmann and Knipping's letter of thanks Haber wrote: "The instrument costs 100,000 marks. Consequently, a claim by the KWI für Physik for joint possession of the instrument would be unfulfillable." (Kallmann and Knipping to KWI für Physik, with addition by Haber, 2 October 1922, AMPG, I. Abt., Rep. 34, Nr. 6, folder Knipping).

[235] Koenigsberger to Einstein, 8 February 1922, AMPG, I. Abt., Rep. 34, Nr. 7, folder Koenigsberger.

[236] Minutes of the Direktorium meeting, 18 May 1922, AMPG, I. Abt., Rep. 34, Nr. 12; Einstein to Kuratorium, 12 June 1922, AMPG, I. Abt., Rep. 1 A, Nr. 1658, p. 197.

[237] Ilse Einstein to Kuratorium, 30 November 1922, AMPG, I. Abt., Rep. 1 A, Nr. 1659, p. 11.

[238] On Pringsheim, see Hanle (1956), Hoffmann (2001).

between himself and Weigert (Pringsheim 1922, 1923; Weigert 1922). In the following years, the KWI für Physik repeatedly funded research undertaken by Pringsheim and his co-workers.[239]

The application submitted by Werner Kolhörster, a physics teacher at a Berlin Gymnasium and guest scholar at the Physikalisch-Technische Reichsanstalt, concerned a totally different kind of research. Kolhörster requested an unspecified amount of money in order to continue investigations into "penetrating rays" that he had initiated before the war, as assistant to Wigand at the University of Halle.[240] The "penetrating rays," at the time also known as "Hess's radiation," "altitude radiation," or "ultragamma radiation," later became known as cosmic rays. They consist of very energetic heavy particles entering the Earth's atmosphere from all directions in space, whose origination, except for those coming from the sun, is not yet completely known. In the twenties, the study of cosmic rays was still in its incipient stage.[241] Nernst, in the context of his reflections on cosmological and astrophysical problems, had found in radioactivity a possible cause for solar heat. Tentatively, he related the radioactive processes with the penetrating rays, whose possible sources might be, according to him, young giant stars in the Milky Way or in nebulae (Nernst 1921a, 36, 59). In order to test this hypothesis, Nernst, who in the spring of 1922 had taken over the directorship of the Reichsanstalt from Warburg, encouraged Kolhörster to resume his observations, offered him a position in the Reichsanstalt, and helped him to raise the necessary funds for expeditions (Kolhörster 1977, 34–36; Mendelssohn 1973, 113–114). In June 1922, Kolhörster received from the KWI für Physik 10,000 marks, not only for equipment but also for operating expenses.[242] Because inflation was rapidly reducing the value of money, he was given an additional 100,000 marks in March 1923, and a further two million marks in July 1923, in the following fiscal year.[243] For this funding he duly thanked the KWI, and above all Nernst, in his final report (Kolhörster 1923, 377). After a campaign of measurements taken under water and on Mount Jungfraujoch in Switzerland, Kolhörster came to the conclusion that "the source of the radiation seems to be near the Milky Way or at least to reach a maximum there," (Kolhörster 1923, 377) thus supporting Nernst's hypothesis. In the following years, he continued his research on cosmic radiation with further support from the KWI[244] and became a recognized authority in the field.

---

[239]See, e.g.., minutes of the Direktorium meeting, 15 May 1924, AMPG, I. Abt., Rep. 1 A, Nr. 1660, pp. 107–108; minutes of the Direktorium meeting, 23 July 1925, GStA, I. HA, Rep. 76 76 V c, Sekt. 2, Tit. 23, Litt. A, Nr. 116, pp. 139.

[240]Kolhörster to Einstein, 1 June 1922, AMPG, I. Abt., Rep. 34, Nr. 7, folder Kolhörster. On Kolhörster and his research, see Kolhörster (1977), Flügge (1980), Hergert (1993).

[241]On the state of the art at that time, see Kolhörster (1924), Wigand (1924).

[242]Minutes of the Direktorium meeting, 15 June 1922, AMPG, I. Abt., Rep. 34, Nr. 12; Einstein to Kuratorium, 26 July 1922, AMPG, I. Abt., Rep. 1 A, Nr. 1658, p. 203a.

[243]Kolhörster to Direktorium, 7 March (with note by Nernst) and 26 June 1923, and Laue to the Direktorium members, 2 July 1923, all in AMPG, I. Abt., Rep. 34, Nr. 7, folder Kolhörster; Laue to Kuratorium, 8 March 1923 and 6 July 1923, AMPG, I. Abt., Rep. 1 A, Nr. 1659, pp. 26 and 41 respectively.

[244]Minutes of the Direktorium meeting, 3 July 1924, AMPG, I. Abt., Rep. 1 A, Nr. 1660, p. 131.

In the spring of 1922, around the time that Kolhörster was encouraged by Nernst to resume his research, Egon von Schweidler, ordinary professor of physics and director of the Physical Institute of the University of Innsbruck in Austria, requested the support of the KWI für Physik for "experimental investigations into penetrating electrical, corpuscular radiation of cosmic origin."[245] Schweidler was an expert on radioactivity and atmospheric electricity, and had already done research on the possible origins of "the Hess radiation," coming to the conclusion that this radiation probably originates from matter evenly distributed in the cosmos, a view quite similar to Nernst's.[246] Nonetheless, the Direktorium was not convinced and postponed its decision "because a closer look at his project has to be made,"[247] thus practically rejecting the request. It is not possible to say what the reason for this skepticism was because there are no other documents related to the case, the only project to be refused support in this budgetary period. In the following years as well, Schweidler never received support from the KWI. On the other hand, during the years 1924/26, the KWI für Physik repeatedly financed another investigation concerning cosmic rays initiated by Gerhard Hoffmann, professor at the University of Königsberg.[248] In fact, Hoffmann had been invited by Nernst himself to submit his request to the institute.[249] Presumably, he and Kolhörster enjoyed Nernst's trust, whereas Schweidler did not.

Finally, a large allocation should be reported that related in principle to both Freundlich's and Einstein's interests in testing general relativity and to the collection of spectroscopic data for understanding the quantum processes in atoms. On the surface, Einstein played a role as director of the KWI für Physik, but in reality he was involved only with the administrative paperwork. In spring of 1922, Walter Grotrian, *Privatdozent* for experimental physics at the University of Göttingen and specialist in spectroscopy, was appointed as observer at the Astrophysikalisches Observatorium of Potsdam with the understanding that he would receive extra funding to equip his new laboratory. The intent was that he should work in astrophysics with newly developed physical methods. In the previous years, Grotrian had worked at the Physical Institutes of the Universities of Frankfurt and Göttingen under the direction of Born and Franck, and had also spent a short time at Haber's KWI für physikalische Chemie und Elektrochemi.e. His call to the Observatory was not due to Einstein's intervention, but rather to the support of his former teachers and of Planck.[250] In March 1922,

---

[245] Einstein to the Direktorium members, 14 May 1922, AMPG, I. Abt., Rep. 34, Nr. 12. Schweidler's application is missing.

[246] Schweidler (1915). See also Kolhörster (1924, 65–66).

[247] Minutes of the Direktorium meeting, 18 May 1922, in AMPG, I. Abt., Rep. 34, Nr. 12.

[248] Hoffmann to KWI für Physik, 10 May and 18 December 1924, 11 February 1926; Hoffmann to Laue, 19 December 1924, 28 February 1925; KWI für Physik to Hoffmann, 1 March 1925; Laue to Hoffmann, 19 February 1926; all in AMPG, I. Abt., Rep. 34, Nr. 4, folder Hoffmann; minutes of the Direktorium meeting, 15 May 1924, AMPG, I. Abt., Rep. 1 A, Nr. 1660, pp. 107–108a.

[249] Hoffmann to KWI für Physik, 18 December 1924, AMPG, I. Abt., Rep. 34, Nr. 4, folder Hoffmann.

[250] The documents related to Grotrian's appointment are in GStA, I. HA, Rep. 76 76 V c, Sekt. 1, Tit. 11, Teil 2, Nr. 6 b, Bd. 8, pp. 333–341. See also minutes of the meeting of the Kuratorium of the

at the end of the fiscal year 1921/22, the Kaiser-Wilhelm-Gesellschaft had a balance of 300,000 marks which, in a period of galloping inflation, had to be disposed of very quickly.[251] At the suggestion of Westphal, who had meanwhile become an official of the Prussian Ministry of Education and its representative on some boards of the Gesellschaft, the KWG decided to spend the sum for Grotrian. For statutory reasons, the money had to go through the KWI für Physik. Therefore, Planck took the trouble to write to Einstein, who was lecturing in Paris at the time, that the KWI für Physik had to make a formal request to the KWG. On the other hand, Grotrian should submit his application to the KWI.[252] And so it happened. The KWI's Direktorium gave its formal approval in May, after the return of Einstein, who had even received a phone call from the Ministry concerning the matter. Ironically enough, Einstein had to send to the administration of the KWG a second request with more detailed scientific justifications because his first one lacked an exposition of motivation.[253] He then explained: "It is intended to recruit Grotrian, one of our best spectroscopists, to the Astrophysikalisches Observatorium in Potsdam in order to introduce the application of modern methods of physical research into astronomical problems," and asked the KWG for 300,000 marks for the laboratory equipment.[254] In December 1922, Grotrian applied for additional funding because inflation had devoured the entire sum without him being able to buy all the required apparatuses.[255] Laue, after discussion within the Directorium during Einstein's absence, approved another 100,000 marks.[256] Thus the largest sum during this period went to Grotrian. In the following years, Grotrian repeatedly received grants from the KWI für Physik and used them to buy more than one hundred different instruments.[257] As it turned out, his research did not contribute to the testing of general relativity but produced valuable data about the constitution of atoms and the nature of the solar corona.[258]

Astrophysikalisches Observatorium, 5 [recte 4] March 1922, GStA, I. HA, Rep. 76 76 V c, Sect. 1, Tit. 11, Teil 2, Nr. 6 6h, pp. 12–13.

[251] As to the cost of living, 300,000 marks in March 1922 corresponded to 10,345 marks in 1913, see Laursen and Pedersen (1964, 134).

[252] Planck to Einstein, 5 April 1922, AEA, 19–296; Grotrian to Einstein, 12 May 1922, AEA, 40–145; Grotrian to Director of the KWI für Physik, 12 May 1922, AMPG, I. Abt., Rep. 34, Nr. 3, folder Grotrian.

[253] Einstein to von Harnack, 19 April 1922, AMPG, I. Abt., Rep. 34, Nr. 13; Glum to Einstein, 22 April 1922, ibid.; minutes of the Direktorium meeting, 18 May 1922, AMPG, I. Abt., Rep. 34, Nr. 12; Einstein to Kuratorium, 12 June 1922, AMPG, I. Abt., Rep. 1 A, Nr. 1658, p. 197.

[254] Einstein to Glum, 28 April 1922, AMPG, I. Abt., Rep. 34, Nr. 3, folder Grotrian; see also Schmidt-Ott to Einstein, 5 May 1922, ibid.; von Harnack to Westphal, 5 May 1922, GStA, I. HA, Rep. 76 76 V c, Sect. 2, Tit. 23, Litt. A, Nr. 116, p. 79.

[255] Grotrian to KWI für Physik, 14 December 1922, AMPG, I. Abt., Rep. 34, Nr. 3, folder Grotrian.

[256] Laue to Ilse Einstein, 15 December 1922, ibid.

[257] All the documents relating to Grotrian for the following years are in AMPG, I. Abt., Rep. 34, Nr. 3, folder Grotrian.

[258] On Grotrian and his work, see Kienle (1954), von Klüber (1966).

Annual Report 1922/23

We have been unable to find an annual report of the KWI für Physik similar to those discussed above, signed by the director and covering the period April 1922 to March 1923. Only an incomplete list was found showing sums spent on projects between October 1922 and July 1923, with Grotrian, Kallman and Knipping, Koenigsberger, Kolhörster, Pohl, and Pringsheim as recipients up to March 1923.[259] However, there is an official report of the KWG covering a different period, namely the time from October 1922 to December 1923. The section on the KWI für Physik (von Harnack 1923, 24) reports that a "series of research works" had been supported and mentions at first two undertakings considered to pertain to astronomy, namely an expedition to Mexico "for the further testing of the solar light deflection and the investigation of the solar corona"—in connection with the solar eclipse of September 1923—and Kolhörster's research on cosmic radiation. The expedition, led by the director of Potsdam's Astrophysikalisches Observatorium, Hans Ludendorff, was financed with two million marks granted by the KWI für Physik in May 1923, that is, in the budgetary year 1923/24.[260] The report then goes on: "All the other supported investigations are connected to atomic physics." In addition to Grotrian's new spectroscopic laboratory, funds were dispensed for research with X-rays by Pohl in Göttingen and by Georg Jaeckel and Walther Kutzner, assistants of Baeyer at the Landwirtschaftliche Hochschule in Berlin, as well as for the study of the emission of electrons through the action of α-rays undertaken by August Becker, extraordinary professor of physics at Lenard's institute at the University of Heidelberg. Even Trautz's measurements of the specific heat of gases were labelled "atomic physics." Finally, a fellowship is listed for a collaborator of Sommerfeld in Munich, "for the calculation of spectra." Apart from the appropriations for Grotrian and Pohl, all others were decided after March 1923.[261] The fellowship for Sommerfeld's assistant indicates that a change of policy had taken place, but no discussion on the question is recorded.

The means available during the year amounted to 1,121,523 marks, 918,521 marks of which were expended.[262] At the end of March 1923, there was a balance of 203,002 marks but this sum corresponded, if related to the cost of living, to only 71.13 marks in 1913.[263] A little less than 40,000 marks had been expended on the salaries of the director—augmented from 20,000 to 60,000 marks a year as of July 1922[264]—and his secretary. As to expenditures for scientific purposes, more than half a million marks was provided for research grants, while 366,000 marks were spent for a purpose not covered by the aims and rules of the KWI für Physik and, what is more, without a

---

[259]"Verzeichnis der im Jahre 1923 bewilligten Zuwendungen," AMPG, I. Abt., Rep. 34, Nr. 13.

[260]Von Laue to Kuratorium, 17 May 1923, AMPG, I. Abt., Rep. 1 A, Nr. 1659, p. 29.

[261]Von Laue to Kuratorium, 22 June and 6 July 1923, AMPG, I. Abt., Rep. 1 A, Nr. 1659, pp. 36a and 41a respectively.

[262]Balance sheet 1 April 1922–31 March 1923, AMPG, I. Abt., Rep. 1 A, Nr. 1665, p. 61. The document is signed by Laue and was sent to the KWG in February 1924 (see ibid., p. 65). See also above, Tables 3.1 and 3.2.

[263]Set 1 for 1913, the cost of living index had risen to 2,854 (Laursen and Pedersen 1964, 135).

[264]Von Harnack to Einstein, 30 October 1922, AMPG, I. Abt., Rep. 34, Nr. 2, folder Einstein.

related decision being made by the Direktorium: the sum was given in March 1923 to the Deutsches Entomologisches Museum (German Entomological Museum).[265] Due to inflation, this institution had lost its capital and was taken over by the KWG as of 1 October 1922, under the condition that no extra costs would arise for the KWG (vom Brocke 1990, 247–248). Apparently, this promise could not be kept and the new burden was, probably, counterbalanced by financial transfers from other KW Institutes initiated by the central administration of the KWG.

The way in which the support for Grotrian had been decided, and also the way in which funds were transferred to the Entomological Museum, show not only that the lack of a spending initiative at the KWI für Physik under Einstein's directorship had been noticed by the KWG administration, but also that someone else had started to make decisions in place of the institute's Direktorium, in particular with regard to the distribution of the much needed money. Nernst's support policy also gives the impression that he considered the institute little more than a source of funding.

3.5   Einstein Ends His Activities at the KWI für Physik

Laue Takes over

In the meeting of 3 March 1921, the Direktorium decided to elect Laue as a new member. The KWI für Physik did not give its approval until December and consequently Laue attended his first Direktorium meeting only on 26 January 1922.[266] On 12 July 1922, Einstein informed Laue that in October he would go on a trip "for heaven knows how long" and therefore asked him to act as deputy director of the KWI für Physik from 1 October for an indefinite period of time. Of course, Laue would then receive the director's salary in lieu of Einstein. Einstein had "so to say officially already left" Berlin and would not be present at the Direktorium meeting scheduled for the next day.[267] On the same day, Einstein also wrote to Planck: "I asked Laue to take over the directorship provisionally, on condition of approval by the Direktorium of the KW-Institute for Physics."[268] This arrangement was accepted without further ado by everyone but was at first believed to be a temporary measure lasting only until Einstein's return.[269]

---

[265] Balance sheet 1 April 1922–31 March 1923, AMPG, I. Abt., Rep. 1 A, Nr. 1665, p. 61; bank statement 1 January–31 March 1923, AMPG, I. Abt., Rep. 34, Nr. 8, folder Mendelssohn.

[266] Einstein to Kuratorium, 7 March and 20 October 1921; Schmidt-Ott to Einstein, 2 December 1921, all in AMPG, I. Abt., Rep. 1 A, Nr. 1658, pp. 166–167, 181, and 184 respectively (Kirsten and Treder 1979, 154, vol. 1).

[267] Einstein to Laue, 12 July 1922, AMPG, I. Abt., Rep. 34, Nr. 2, folder Einstein.

[268] Einstein to Planck, 12 July 1922, AEA, 19–304.

[269] Schlüter (1994, 97–99; 1995, 184–185); Laue to von Harnack, 14 October 1922, AMPG, I. Abt., Rep. 1 A, Nr. 1659, p. 5; von Harnack to Laue, 16 November 1922, AMPG, I. Abt., Rep. 34, Nr. 8, folder Laue.

In fact, after his return to Berlin in March 1923, Einstein turned over the administration of the institute to Laue although he kept the formal position of director of the KWI für Physik. In the following years he occasionally attended the meetings of the Direktorium, but we do not have any document signed by him in his function as director after October 1922. We can therefore assert that Einstein essentially stepped down from his managing position. However, he continued to function as a figurehead for the institute. The official publications of the KWG as well as its annual reports, published in the journal *Die Naturwissenschaften*, continue to list Einstein as director and Laue as deputy director until 1932.[270] As Laue explained: "On the one hand, I am deputy director although Einstein's directorship is used only as a decoration. On the other hand, grants are decided by the Direktorium as a whole."[271] After Laue's entrance into the Direktorium, administrative activities increased immediately. As already noted for 1922, the Direktorium meetings took place more frequently,[272] the minutes of the meetings became more detailed, and a new secretary replaced Einstein's stepdaughter.[273] Whether the funding policy also changed is yet to be studied.

Why Did Einstein Give Up His Managing Directorship?

In the summer of 1922, Einstein decided that he would withdraw from the managing of the KWI für Physik. What reasons did he have? Officially, Einstein justified the decision with his absence from Berlin for an indefinite time. In fact, he planned to accept an invitation to lecture in Japan and would thus be away for several months. He wanted to leave Germany for a time in order to escape public attention because he felt his life was in danger (Seelig 1960, 305–307; Fölsing 1993, 594–597). Since the end of 1919—that is, since the results of the famous British expeditions had been made known, which seemingly verified his predictions concerning the deflection of solar light—Einstein, to his discomfort (Fölsing 1993, 504) was at the center of public interest: "Guilty for the whole mess is in the end the English expedition of 1919 [...] Since then I have become a kind of flag."[274] In 1920 he had become the target of public and, to a great extent, politically motivated attacks in the press and in meetings.[275] In June 1922, the German Minister for Foreign Affairs, Walther Rathenau, had been murdered by anti-semitic nationalists, and Einstein, who was already known as a pacifist and internationalist, was also in great danger: "I received trustworthy information that local nationalist circles are attempting at my life [...].

---

[270]See, e.g., Harnack (1928, 180).

[271]Laue to Hoffmann, 30 January 1926, AMPG, I. Abt., Rep. 34, Nr. 4, folder Hoffmann.

[272]E.g., on 15 May, 14 June, 5 July, 26 July, and 1 November 1923 (see von Laue to Kuratorium, 17 May, 14 June, 6 July, and 27 July 1923, AMPG, I. Abt., Rep. 1 A, Nr. 1659, pp. 29, 35, 41a, and 50 respectively; von Laue to Kuratorium, 6 November 1923, AMPG, I. Abt., Rep. 1 A, Nr. 1660, p. 60).

[273]Hildegard Bathe was engaged as the new secretary starting from 1 July 1923 (Laue to Kuratorium, 16 June 1923, AMPG, I. Abt., Rep. 1 A, Nr. 1659, p. 43).

[274]Einstein to Planck, 12 July 1922, AEA, 19–304.

[275]Fölsing (1993, 519–522); on the controversy over relativity and the campaign against Einstein, see Goenner (1993), Wazeck (2009).

However, being a little fond of my life, I decided to secure my corpse at once."[276] He then canceled his course at the University and his promised appearance at the meeting of the Gesellschaft Deutscher Naturforscher und Aerzte to be held in Leipzig in September. In fact, on 5 July 1922 Einstein left Berlin for Kiel where he stayed for a couple of days.[277] For a while, longing for a "normal human life in peace and quiet,"[278] he pondered in earnest withdrawing from the Academy and leaving Berlin for good, but he finally came back, having decided to give up only his "official charges."[279] The long journey abroad was an ideal pretext.

Thus, at first sight, we could suppose that Einstein resigned from the directorship of the KWI für Physik for political reasons: he no longer wanted to play an official role, at least for the time being. But was this the real reason? We think not. First of all, already on 1 August 1922 Einstein took part in a pacifist demonstration in Berlin (Nathan and Norden 1968, 54). Moreover, his function as director of the institute did not require a public presence. And, anyway, after he returned from his extended trip in 1923, Einstein was still meeting the other Direktorium members almost every Wednesday at the "physical colloquium" of the University and almost every Thursday at the sessions of the Academy. Most of the meetings of the Direktorium of the KWI für Physik took place after the Academy sessions.[280]

We suggest that the main reasons for Einstein's resignation were his discontent with the burdens the administration of the institute entailed and with the ongoing practice of research financing adopted by the Direktorium. We also suggest that his decision was made possible by his attainment of private financial independence. While Einstein was tired of his administrative duties, Laue was willing and able to take on such tasks and enjoyed the trust of the physics community: in March 1922 he had been elected chairman of the newly constituted steering committee at the Astrophysikalisches Observatorium[281] and, in the following May, chairman of the sectional committee on physics of the Notgemeinschaft.[282]

Complaints about the bureaucratic duties to be carried out for the KWI für Physik are recurrent in Einstein's correspondence. Immediately after the start of the institute

---

[276]Einstein to Planck, undated [6 July 1922], AEA, 19–300; see also Nathan and Norden (1968, 53–54). For an account of the episode as well as of Einstein's reaction, see Fölsing (1993, 595–597).

[277]Einstein to Anschütz-Kämpfe, 1 July 1922, in Lohmeier and Schell (1992, 167).

[278]Einstein to Anschütz-Kämpfe, 12 July 1922, in Lohmeier and Schell (1992, 171).

[279]Elsa Einstein to Anschütz-Kämpfe, 16 July 1922, in Lohmeier and Schell (1992, 175).

[280]See, e.g., Laue to the Direktorium members, 11 June 1923, AMPG, I. Abt., Rep. 34, Nr. 7, folder Koch; Laue to the Direktorium members, 26 June and 29 October 1923, ibid., Nr. 1, folder Becker; Laue to the Direktorium members, 29 January 1924, AEA, 40–153; Laue to Glum, 3 April 1924, AMPG, I. Abt., Rep. 1 A, Nr. 1660, p. 96; Laue to Schmidt-Ott, 29 June 1924, ibid., p. 124; minutes of the Direktorium meeting, 26 February 1925, GStA, I. HA, Rep. 76 76 V c, Sekt. 2, Tit. 23, Litt. A, Nr. 116, p. 130.

[281]Minutes of the meeting of the Kuratorium of the Astrophysikalisches Observatorium, 5 [recte 4] March 1922, GStA, I. HA, Rep. 76 76 V c, Sect. 1, Tit. 11, Teil 2, Nr. 6 6h, pp. 12–13.

[282]Laue to Schmidt-Ott, 18 May 1922, AMPG, I. Abt., Rep. 34, Nr. 9, folder Notgemeinschaft; (Richter 1972, 16).

he wrote: "The K. W. Institute involves quite a large amount of correspondence."[283] Shortly thereafter, he expressed disappointment because "nothing worth mentioning" had arrived in response to the KWI's public call for research proposals.[284] He felt like a "slave of the damned postman"[285] wasting his time: "At the moment, my [scientific] work is not up to much. I'm dissipating my forces, I have to clear a huge amount of correspondence, evaluate, advise, patronize, but I'm not progressing on the big problems."[286] He had no understanding for the necessity of bureaucratic procedures. We have already reported that Einstein only bothered to write a report on the institute's activities and to set up a budget after a complaint from the KWI für Physik chairman. It seems that he would actually have preferred to avoid the Direktorium meetings altogether: "The K.W.I. is quite sluggish, since I have always to round up all my dear …for one grant."[287] One of Einstein's closest friends commented at the very moment of his withdrawal: "Einstein is tired of Berlin with all that it implies concerning visits and official matters and, horribile scriptu, wants to change over to technology."[288]

We have reason to believe that Einstein, at least during the last period of his directorship, was very discontent with the funding policy of his institute. In the summer of 1922, Einstein learned from the German-American physiologist, Jacques Loeb, a member of the Rockefeller Institute for Medical Research, that the Rockefeller Foundation intended to support educational institutions and scientific research in Germany. Einstein wrote to Loeb that he hoped that "you will not give your money to organizations, which are bound to thousands of considerations, but directly to those who are able to make something worthwhile of it. Otherwise the money will be spread in little ineffective portions to support superfluous mediocrity."[289] It seems that the question really troubled him because a little later he wrote again: "If you want to give something for the sake of science in Germany, absolutely do not give it to organizations, which always take their decisions according to the principle of minimal odium, but give it in accordance with your free judgment to a few capable people."[290] Clearly, in writing this Einstein had his own institute in mind.

On the other hand, Einstein would not have been so foolish as to give up a considerable salary without compensating revenue, especially since he had to maintain two households.[291] As of April 1922 Einstein's salary at the KWI für Physik had been raised to 18,000 marks net.[292] In addition, he still received a special salary from the

---

[283]Einstein to Besso, 5 January 1918, in Einstein (1998, 598, doc. 428).

[284]"nichts Vernünftiges" (Albert Einstein to Ilse Einstein, 12 May 1918, in Einstein (1998, 758, doc. 536).

[285]Einstein to Born, 22 August 1921, in Einstein and Born (1969, 85).

[286]Einstein to Besso, 26 July 1920, in Einstein and Besso (1972, 152–153).

[287]Einstein to Born, 22 August 1921, in Einstein and Born (1969, 85). Dots by Einstein.

[288]Anschütz-Kaempfe to Sommerfeld, 12 July 1922, in Lohmeier and Schell (1992, 170).

[289]Einstein to Loeb, 14 August 1922, AEA, 15–192.

[290]Einstein to Loeb, 22 September 1922, AEA, 15–193.

[291]His first wife Mileva with two sons in Zurich and his second wife Elsa with two stepdaughters in Berlin.

[292]Von Harnack to Einstein, 28 March 1922, AMPG, I. Abt., Rep. 34, Nr. 2, folder Einstein.

Academy which, since April 1922, had also been raised to 75,000 marks per year.[293] But at least by 1920 his other sources of income were quite remarkable, as is already apparent from a still only preliminary collation of documental sources. For example, Einstein received 2,000 marks for a single talk given in Kiel in September 1920.[294] A few months later, Sommerfeld offered him the same sum for a lecture in Munich: "From Anschütz I hear that you have obtained 2,000 marks in Kiel; we can give you this as well."[295] In January 1921, the industrialist and inventor of the gyrocompass, Hermann Anschütz-Kaempfe, must have paid to Einstein under the table 20,000 marks for his contribution to the development of a new model of gyrocompass: "I write explicitly hand over and not transfer, otherwise inquiries concerning taxes will come up."[296] This payment was followed by others.[297] In a couple of weeks, Einstein earned from these "other" activities a little less than half of his annual academic salaries, which at the time (end of 1920) amounted to 10,000 marks from the KWI and 36,000 marks from the Academy.[298] One talk alone amounted to twenty per cent of his annual director's salary. In addition, Einstein was a very successful author. His "popular account" of the theory of relativity (Einstein 1917a; Gutfreund and Renn 2015) went through fourteen editions from 1917 to 1922, a total of 65,000 copies, more than 45,000 of which were sold in 1920 alone for a price between 2.80 and 4 marks (Oberschelp and Gorzny 1976–1981, 240, vol. 31 and 588, vol. 138; Hermann 1994, 250–251). For this book Einstein received twenty per cent of the selling price, (Fölsing 1993, 558) which means more than 25,000 marks in royalties in just one year.

In summer 1922, Einstein probably earned as much or even more money from his books, public lectures, and from his work on the gyrocompass than he received from the Academy and the KWI: "Materially, I am rather independent, to the extent that the remuneration I get from the Academy is practically negligible so that I can renounce it without disturbing the equilibrium."[299] We thus conclude that Einstein was tired of doing all the administrative work for the institute with little benefit to his own research and felt financially independent enough to withdraw.

---

[293] This was the basic salary of an ordinary university professor; in addition Einstein also received a "local bonus" and an allowance for children (G. Roethe to Einstein, 20 October 1921, and Krüss to Prussian Academy of Sciences, 6 June 1922, AAdW, II–III, Bd. 39, pp. 37 and 108 respectively).

[294] The talk was on "Space and time in the light of relativity theory" (Lohmeier and Schell 1992, 117).

[295] Sommerfeld to Einstein, 29 December 1920, in Einstein and Sommerfeld (1968, 76).

[296] Anschütz-Kaempfe to Einstein, 28 December 1920, in Lohmeier and Schell (1992, 115).

[297] In the correspondence with Anschütz-Kaempfe there are indications that, from the end of 1921, the money coming from Kiel was regularly transferred to Einstein's children in Switzerland. In October 1926, a contract was arranged between Anschütz-Kaempfe's firm and Einstein so that he would receive one percent of the selling price for each gyrocompass and three percent of the license fees should foreign patents be issued (Lohmeier and Schell 1992, 56, 58).

[298] Schmidt-Ott to Einstein, 23 June 1920, AMPG, I. Abt., Rep. 34, Nr. 13; Prussian Ministry of Education to Prussian Academy of Sciences, 12 November 1920, AAdW, II–III, Bd. 38, p. 171.

[299] Einstein to Loeb, 22 September 1922, AEA, 15–193.

References

Bartel, Hans-Georg and Rudolf P. Huebener (2007). *Walther Nernst. Pioneer of Physics and Chemistry*. Singapore: World Scientific.

Bergmann, Ludwig (1958). "Clemens Schaefer zum 80. Geburtstag". In: *Die Naturwissenschaften* 45, pp. 121–122.

Bohlin, Helge (1920). "Eine neue Anordnung für röntgenkristallographische Untersuchungen von Kristallpulver". In: *Annalen der Physik* 61, pp. 421–439.

Brauns, Reinhard (1934). "Kristalle, flüssige". In: *Handwörterbuch der Naturwissenschaften*. Vol. 5. Ed. by Rudolf Dittler, Georg Joos, and Eugen Korschelt. 2nd ed. Jena: Fischer, pp. 1159–1179.

Bucky, Peter A. (1991). *Der private Albert Einstein. Gespräche über Gott, die Menschen und die Bombe*. Düsseldorf: Econ.

Conrad, Richard (1926a). "Über die Streuungsabsorption von Wasserstoffkanalstrahlen beim Durchgang durch Wasserstoff". In: *Zeitschrift für Physik* 38, pp. 465–474.

Conrad, Richard (1926b). "Über die Streuungsabsorption von Wasserstoffkanalstrahlen beim Durchgang durch Wasserstoff und Helium". In: *Zeitschrift für Physik* 35, pp. 73–99.

Debye, Peter and Paul Scherrer (1918). "Atombau". In: *Nachrichten von der Königlichen Gesellschaft der Wissenschaften zu Göttingen: Mathematisch-physikalische Klasse*, pp. 101–120.

Dreisigacker, Ernst and Helmut Rechenberg (1995). "Karl Scheel, Ernst Brüche und die Publikationsorgane". In: *150 Jahre Deutsche Physikalische Gesellschaft*. Ed. by Theo Mayer-Kuckuk. Weinheim: VCH Verlagsgesellschaft. F 135–F 142.

Earman, John and Clark Glymour (1980). "The Gravitational Red Shift as a Test of General Relativity: History and Analysis". In: *Studies in History and Philosophy of Science* 11, pp. 175–214.

Einstein, Albert (1917a). "Kosmologische Betrachtungen zur allgemeinen Relativitätstheorie". In: *Sitzungsberichte der Königlich Preussischen Akademie der Wissenschaften*, pp. 142–152. Reprinted in (Einstein 1996, doc. 43).

Einstein, Albert (1917b). *Über die spezielle und die allgemeine Relativitätstheorie (Gemeinverständlich)*. Braunschweig: Vieweg. Reprinted in (Einstein 1996, doc. 42).

Einstein, Albert (1918). "Der Energiesatz in der allgemeinen Relativitätstheorie". In: *Sitzungsberichte der Königlich Preussischen Akademie der Wissenschaften*, pp. 448–459. Reprinted in (Einstein 2002, doc. 9).

Einstein, Albert (1921). "Über ein den Elementarprozeß der Lichtemission betreffendes Experiment". In: *Sitzungsberichte der Preussischen Akademie der Wissenschaften*, pp. 882–883. Reprinted in (Einstein 2002, doc. 68).

Einstein, Albert (1922a). "Emil Warburg als Forscher". In: *Die Naturwissenschaften* 10, pp. 823–828.

Einstein, Albert (1922b). "Theoretische Bemerkungen zur Supraleitung der Metalle". In: *Het Natuurkundig Laboratorium der Rijksuniversiteit te Leiden in de jaren 1904–1922: Gedenkboek aangeboden aan H. Kamerlingh Onnes*. Leiden: Ijdo, pp. 429–435.

Einstein, Albert (1922c). "Zur Theorie der Lichtfortpflanzung in dispergierenden Medien". In: *Sitzungsberichte der Preussischen Akademie der Wissenschaften: Physikalisch-Mathematische Klasse*, pp. 18–22.

Einstein, Albert (1989). *The Collected Papers*. Vol. 2: *The Swiss Years: Writings, 1900–1909*. Ed. by John Stachel. Princeton: Princeton University Press.

Einstein, Albert (1993a). *The Collected Papers*. Vol. 3: *The Swiss Years: Writings, 1909–1911*. Ed. by Martin J. Klein et al. Princeton: Princeton University Press.

Einstein, Albert (1993b). *The Collected Papers*. Vol. 5: *The Swiss Years: Correspondence, 1902–1914*. Ed. by Martin J. Klein, Anne J. Kox, and Robert Schulmann. Princeton: Princeton University Press.

Einstein, Albert (1998). *The Collected Papers*. Vol. 8: *The Berlin Years: Correspondence, 1914–1918*. Princeton: Princeton University Press.

Einstein, Albert (2002). *The Collected Papers*. Vol. 7: *The Berlin Years: Writings, 1918–1921*. Ed. by Michel Janssen et al. Princeton: Princeton University Press.

Einstein, Albert (2004). *The Collected Papers*. Vol. 9: *The Berlin Years: Correspondence, January 1919–April 1920*. Ed. by Diana Kormos Buchwald et al. Princeton: Princeton University Press.

Einstein, Albert (2006). *The Collected Papers*. Vol. 10: *The Berlin Years: Correspondence, May–December 1920 and Supplementary Correspondence, 1909–1920*. Ed. by Diana Kormos Buchwald et al. Princeton: Princeton University Press.

Einstein, Albert (2009). *The Collected Papers*. Vol. 12: *The Berlin Years: Correspondence, January–December 1921*. Ed. by Diana Kormos Buchwald et al. Princeton: Princeton University Press.

Einstein, Albert and Michele Besso (1972). *Correspondance 1903–1955*. Paris: Hermann. Translated and edited by Pierre Speziali.

Einstein, Albert and Max Born (1969). *Briefwechsel 1916–1955. Kommentiert von Max Born*. Munich: Nymphenburger.

Einstein, Albert and Jakob Grommer (1923). "Beweis der Nichtexistenz eines überall regulären zentrisch symmetrischen Feldes nach der Feld-Theorie von Th. Kaluza". In: *Scripta Universitatis atque Bibliothecae Hierosolymitanarum: Mathematica et Physica* 1. VII.

Einstein, Albert and Arnold Sommerfeld. (1968). *Briefwechsel. Sechzig Briefe aus dem goldenen Zeitalter der modernen Physik*. Ed. by Armin Hermann. Basel: Schwabe. Herausgegeben und kommentiert von Armin Hermann.

Estermann, Immanuel (1976). "Stern, Otto". In: *Dictionary of Scientific Biography*. Vol. 13. Ed. by Charles C. Gillispie. New York: Scribner, pp. 40–43.

Flügge, Siegfried (1980). "Kolhörster, Werner". In: *Neue Deutsche Biographie*. Vol. 12. Berlin: Duncker & Humblot, pp. 460–461.

Fölsing, Albrecht (1993). *Albert Einstein. Eine Biographie*. Frankfurt am Main: Suhrkamp.

Forman, Paul (1968). *The Environment and Practice of Atomic Physics in Weimar Germany*. Ann Arbor (Michigan): UMI.

Försterling, Karl and G. Hansen (1923). "Zeemaneffekt der roten und blauen Wasserstofflinie". In: *Zeitschrift für Physik* 18, pp. 26–33.

Fouquet, Dörte (1999). *Die Gründung der Hamburgischen Universität*. Potsdam: Verlag für Berlin-Brandenburg.

Franck, James (1921). "Über Lichtanregung und Ionisation von Atomen und Molekülen durch Stöße langsamer Elektronen". In: *Physikalische Zeitschrift* 22, pp. 388–391, 409–414, 441–448, 466–471.

Franck, James and Walter Grotrian (1921). "Bemerkungen über angeregte Atome". In: *Zeitschrift für Physik* 4, pp. 89–99.

Franck, James and Paul Knipping (1919). "Die Ionisierung des Helium". In: *Physikalische Zeitschrift* 21, pp. 481–488.

Franck, James and Paul Knipping (1920). "Über die Anregungsspannungen des Heliums". In: *Zeitschrift für Physik* 1, pp. 320–332.

Füchtbauer, Christian and Georg Joos (1922). "Über Intensität und Verbreiterung von Spektrallinien". In: *Physikalische Zeitschrift* 23, pp. 73–80.

Füchtbauer, Christian, Georg Joos, and Otto Dinckelacker (1923). "Über Intensität, Verbreiterung und Druckverschiebung von Spektrallinien, insbesondere der Absorptionslinie 2537 des Quecksilbers". In: *Annalen der Physik* 71, pp. 204–227.

Fürth, Reinhold (1920). "Bemerkungen zu Herrn E. Radels Arbeit: Ladungsbestimmungen an Nebelteilchen". In: *Zeitschrift für Physik* 3, pp. 422–424.

Gerlach, Walther (1978). "Robert Wichard Pohl". In: *Jahrbuch der Bayerischen Akademie der Wissenschaften*, pp. 214–219.

Gerlach, Walther and Otto Stern (1922a). "Das magnetische Moment des Silberatoms". In: *Zeitschrift für Physik* 9, pp. 353–355.

Gerlach, Walther and Otto Stern (1922b). "Der experimentelle Nachweis der Richtungsquantelung im Magnetfeld". In: *Zeitschrift für Physik* 9, pp. 349–352.

Gerlach, Walther and Otto Stern (1922c). "Der experimentelle Nachweis des magnetischen Moments des Silberatoms". In: *Zeitschrift für Physik* 8, pp. 110–111.

Gerlach, Walther and Otto Stern (1924). "Über die Richtungsquantelung im Magnetfeld". In: *Annalen der Physik* 74, pp. 673–699.

Goenner, Hubert (1993). "The Reaction to Relativity Theory I: The Anti-Einstein Campaign in Germany in 1920". In: *Science in Context* 6, pp. 107–133.

Goenner, Hubert (2004). "On the History of Unified Field Theories". In: *Living Reviews in Relativity* 7.2. https://link.springer.com/article/10.12942/lrr-2004-2 (accessed 12/21/2020).

Goenner, Hubert and Giuseppe Castagnetti (1996). "Albert Einstein as Pacifist and Democrat during World War I". In: *Science in Context* 9, pp. 325–386.

Grebe, Leonhard (1920). "Über die Gravitationsverschiebung der Fraunhoferschen Linien". In: *Physikalische Zeitschrift* 21, pp. 662–668.

Grebe, Leonhard (1921). "Sonnengravitation und Rotverschiebung". In: *Zeitschrift für Physik* 4, pp. 105–109.

Grebe, Leonhard and Albert Bachem (1919). "Über den Einsteineffekt im Gravitationsfeld der Sonne". In: *Verhandlungen der Deutschen Physikalischen Gesellschaft* 21, pp. 454–464.

Grebe, Leonhard and Albert Bachem (1920a). "Die Einsteinsche Graviationsverschiebung im Sonnenspektrum der Stickstoffbande $\lambda = 3883$ AE". In: *Zeitschrift für Physik* 2, pp. 415–422.

Grebe, Leonhard and Albert Bachem (1920b). "Über die Einsteinverschiebung im Graviationsfeld der Sonne". In: *Zeitschrift für Physik* 1, pp. 51–54.

Grommer, Jakob (1919). "Beitrag zum Energiesatz in der allgemeinen Relativitätstheorie". In: *Sitzungsberichte der Preussischen Akademie der Wissenschaften*, pp. 860–862.

Gudden, Bernhard (1944). "R. W. Pohl zum 60. Geburtstag". In: *Die Naturwissenschaften* 32, pp. 166–169.

Gutfreund, Hanoch, and Jürgen Renn, eds. (2015). *Relativity: The Special and the General Theory, 100th Anniversary Edition.* Princeton: Princeton University Press.

Günther, Paul (1920). "Über die innere Reibung des Wasserstoffs bei tiefen Temperaturen". In: *Sitzungsberichte der Preussischen Akademie der Wissenschaften*, pp. 720–726.

Günther, Paul (1924a). "Die kosmologischen Betrachtungen von Nernst". In: *Zeitschrift für angewandte Chemie* 37, pp. 454–457.

Günther, Paul (1924b). "Über die innere Reibung der Gase bei tiefen Temperaturen". In: *Zeitschrift für physikalische Chemie* 110, pp. 626–636.

Hammer, Wilhelm (1919–1920). "Die Messung kleiner Kapazitäts- und Selbstinduktions- Änderungen mittels ungedämpfter Schwingungen". In: *Berichte der naturforschenden Gesellschaft zu Freiburg im Breisgau* 22.2, pp. 1–6.

Hanle, Wilhelm (1956). "Peter Pringsheim 75 Jahre". In: *Physikalische Blätter* 12, pp. 126–127.

Hentschel, Klaus (1990). *Interpretationen und Fehlinterpretationen der speziellen und der allgemeinen Relativitätstheorie durch Zeitgenossen Albert Einsteins.* Basel: Birkhäuser.

Hentschel, Klaus (1992a). *Der Einstein-Turm: Erwin F. Freundlich und die Relativitätstheorie.* Heidelberg: Spektrum Akademischer Verlag.

Hentschel, Klaus (1992b). "Einstein's Attitude towards Experiments: Testing Relativity Theory 1907–1927". In: *Studies in History and Philosophy of Science* 23, pp. 593–624.

Hentschel, Klaus (1992c). "Grebe/Bachems photometrische Analyse der Linienprofile und die Gravitations-Rotverschiebung: 1919 bis 1922". In: *Annals of Science* 49, pp. 21–46.

Hentschel, Klaus (1998). *Zum Zusammenspiel von Instrument, Experiment und Theorie. Rotverschiebung im Sonnenspektrum und verwandte spektrale Verschiebungseffekte von 1880 bis 1960.* Hamburg: Kovac.

Hergert, Wolfram (1993). "Physik im Freiballon. Forschungen zur Höhenstrahlung und zur Physik der Atmosphäre am Physikalischen Institut der Universität Halle in den Jahren 1910–1937". In: *Physikalische Blätter* 49, pp. 1007–1010.

Hermann, Armin (1994). *Einstein. Der Weltweise und sein Jahrhundert.* Munich: Piper.

Hermann, Armin (1995). "Die Deutsche Physikalische Gesellschaft 1899–1945". In: *150 Jahre Deutsche Physikalische Gesellschaft.* Ed. by Theo Mayer-Kuckuk. Weinheim: VCH Verlagsgesellschaft. F 61–F 105.

Hettner, Gerhard (1920a). "Über den Einfluß eines äußeren elektrischen Feldes auf das Rotationsspektrum. Ein Analogon zum Starkeffekt". In: *Zeitschrift für Physik* 2, pp. 349–360.

Hettner, Gerhard (1920b). "Über Gesetzmäßigkeiten in den ultraroten Gasspektren und ihre Deutung". In: *Zeitschrift für Physik* 1, pp. 345–354.

Hettner, Gerhard (1922). "Die Bedeutung von Rubens Arbeiten für die Plancksche Strahlungsformel". In: *Die Naturwissenschaften* 10, pp. 1033–1038.

Hiebert, Erwin N. (1978). "Nernst, Hermann Walther". In: *Dictionary of Scientific Biography*. Ed. by Charles C. Gillispie. Vol. 15, Supplement I. New York: Scribner, pp. 432–453.

Hoffmann, Dieter (1986). "Pionier der Plasmaphysik. Zum 100. Geburtstag des Greifswalder Physikers Rudolf Seeliger". In: *Wissenschaft und Fortschritt* 36, pp. 278–280.

Hoffmann, Dieter (2001). "Pringsheim, Peter". In: *Neue Deutsche Biographie*. Vol. 20. Berlin: Duncker & Humblot, pp. 725–726.

Hoffmann, Dieter and Edgar Swinne (1994). *Über die Geschichte der "technischen Physik" in Deutschland und den Begründer ihrer wissenschaftlichen Gesellschaft Georg Gehlhoff*. Berlin: ERS-Verlag.

Holton, Gerald (1978). "Subelectrons, Presuppositions, and the Millikan-Ehrenhaft Dispute". In: *Historical Studies in the Physical Sciences* 9, pp. 161–224.

Hoyer, Ulrich (1993). "Walther Gerlach (1889–1979)". In: *Naturwissenschaften und Technik in der Geschichte. 25 Jahre Lehrstuhl für Geschichte der Naturwissenschaft und Technik am Historischen Institut der Universität Stuttgart*. Ed. by Helmuth Albrecht. Stuttgart: Verlag für Geschichte der Naturwissenschaften und der Technik, pp. 133–141.

Hund, Friederich, Heinz Maier-Leibnitz, and Erich Mollwo (1988). "Physics in Göttingen—with Franck, Born and Pohl". In: *European Journal of Physics* 9, pp. 188–194.

Jungnickel, Christa and Russell McCormmach (1986). *Intellectual Mastery of Nature. Theoretical Physics from Ohm to Einstein*. Chicago: The University of Chicago Press.

Kangro, Hans (1970). *Vorgeschichte des Planckschen Strahlungsgesetzes*. Wiesbaden: Steiner.

Kant, Horst (1987). "Das KWI für Physik—von der Gründung bis zum Institutsbau". In: *Berliner Wissenschaftshistorische Kolloquien XII: Beiträge zur Astronomie- und Physikgeschichte*. Berlin: Akademie der Wissenschaften der DDR. Institut für Theorie, Geschichte und Organisation der Wissenschaft, pp. 129–141.

Kant, Horst (1993). "Peter Debye und das Kaiser-Wilhelm-Institut für Physik in Berlin". In: *Naturwissenschaften und Technik in der Geschichte. 25 Jahre Lehrstuhl für Geschichte der Naturwissenschaft und Technik am Historischen Institut der Universität Stuttgart*. Ed. by Helmuth Albrecht. Stuttgart: Verlag für Geschichte der Naturwissenschaften und der Technik, pp. 161–177.

Kaufmann, Walther and Fritz Serowy (1921). "Druckmessung mittels Glühkathodenröhren". In: *Zeitschrift für Physik* 5, pp. 319–323.

Kienle, Hans (1954). "Nachruf auf Walter Grotrian". In: *Jahrbuch der Deutschen Akademie der Wissenschaften zu Berlin*, pp. 370–374.

Kirschbaum, Heinrich (1923). "Über die Intensitätsverteilung und den Ursprung des Bandenspektrums von Stickstoff". In: *Annalen der Physik* 71, pp. 289–316.

Kirsten, Christa and Hans-Jürgen Treder, eds. (1979). *Albert Einstein in Berlin 1913–1933*, 2 vols. Berlin: Akademie-Verlag.

Knipping, Paul (1921). "Die Ionisierungsspannungen der Halogenwasserstoffe". In: *Zeitschrift für Physik* 7, pp. 328–340.

Knipping, Paul (1923). "Registrierapparat zur automatischen Aufnahme von Ionisierungs- und anderen Kurven". In: *Zeitschrift für Instrumentenkunde* 43, pp. 241–256.

Koch, Peter P. (1912). "Über ein registrierendes Mikrophotometer". In: *Annalen der Physik* 39, pp. 705–751.

Kolhörster, Editha (1977). *Mein Leben an der Seite eines Wissenschaftlers und Forschers*. Darmstadt-Eberstadt: Kolhörster.

Kolhörster, Werner (1923). "Intensitäts- und Richtungsmessungen der durchdringenden Strahlung". In: *Sitzungsberichte der Preussischen Akademie der Wissenschaften: Physikalisch-mathematische Klasse*, pp. 366–377.

Kolhörster, Werner (1924). *Die durchdringende Strahlung in der Atmosphäre*. Hamburg: Grand.

Kolhörster, Werner (1933). "Albert Wigand zum Gedenken". In: *Meteorologische Zeitschrift* 50, pp. 60–62.

Kulenkampff, Helmuth (1922). "Über das kontinuierliche Röntgenspektrum". In: *Annalen der Physik* 69, pp. 548–596.

L'Autographe (1992). *Lettres autographes. Documents historiques. Photos signées*. Catalogue 24. Geneva.

Langevin, Paul and Maurice de Broglie, eds. (1912). *La théorie du rayonnement et les quanta. Rapports et discussions de la Réunion tenue à Bruxelles, du 30 octobre au 3 novembre 1911, sous les Auspices de M. E. Solvay*. Paris: Gauthier-Villars.

Laue, Max and Friedrich Franz Martens (1907). "Bestimmung der optischen Konstanten von glühenden Metallen aus der Polarisation der seitlich emittierten Strahlung". In: *Physikalische Zeitschrift* 8, pp. 853–856.

Laursen, Karsten and Jørgen Pedersen (1964). *The German Inflation 1918–1923*. Amsterdam: North-Holland.

Lohmeier, Dieter and Bernhardt Schell, eds. (1992). *Einstein, Anschütz und der Kieler Kreiselkompaß. Der Briefwechsel zwischen Albert Einstein und Hermann Anschütz-Kaempfe und andere Dokumente*. Heide in Holstein: Westholsteinische Verlagsanstalt Boyens.

Marsch, Ulrich (1994). *Notgemeinschaft der Deutschen Wissenschaft. Gründung und frühe Geschichte 1920–1925*. Frankfurt am Main: Lang.

Mehra, Jagdisch and Helmut Rechenberg (1982). *The Historical Development of Quantum Theory*. Vol. 1: *The Quantum Theory of Planck, Einstein, Bohr and Sommerfeld: Its Foundation and the Rise of Its Difficulties 1900–1925*. New York: Springer.

Mendelssohn, Kurt (1973). *The World of Walther Nernst. The Rise and Fall of German Science 1864–1941*. London: Macmillan Press.

Müller, Gustav (1919). "[Jahresberichte der Sternwarten für 1918] Potsdam (Astrophysikalisches Observatorium)". In: *Vierteljahrsschrift der Astronomischen Gesellschaft* 54, pp. 246–252.

Müller, Gustav (1920). "[Jahresberichte der Sternwarten für 1919] Potsdam (Astrophysikalisches Observatorium)". In: *Vierteljahrsschrift der Astronomischen Gesellschaft* 55, pp. 112–117.

Müller, Gustav (1921). "[Jahresberichte der Sternwarten für 1920] Potsdam (Astrophysikalisches Observatorium)". In: *Vierteljahrsschrift der Astronomischen Gesellschaft* 56, pp. 114–119.

Nathan, Otto and Heinz Norden (1968). *Einstein on Peace*. New York: Schoken.

Nernst, Walther (1919). "Einige Folgerungen aus der sogenannten Entartungstheorie der Gase". In: *Sitzungsberichte der Preussischen Akademie der Wissenschaften*, pp. 118–127.

Nernst, Walther (1921a). *Das Weltgebäude im Lichte der neueren Forschung*. Berlin: Springer.

Nernst, Walther (1921b). *Theoretische Chemie vom Standpunkte der Avogadroschen Regel und der Thermodynamik*. 8th–10th ed. Stuttgart: Enke.

Nernst, Walther and Theodor Wulf (1919). "Über eine Modifikation der Planckschen Strahlungsformel auf experimenteller Grundlage". In: *Verhandlungen der Deutschen Physikalischen Gesellschaft* 21, pp. 294–337.

Oberschelp, Reinhard and Willi Gorzny, eds. (1976–1981). *Gesamtverzeichnis des deutschsprachigen Schrifttums (GV): 1911–1965*. Munich: Saur.

Pais, Abraham (1982). *"Subtle is the Lord . . . " The Science and the Life of Albert Einstein*. Oxford: Oxford University Press.

Perrin, Jean (1912). "Les preuves de la réalité moléculaire (étude spéciale des émulsions)". In: *La théorie du rayonnement et les quanta. Rapports et discussions de la Réunion tenue à Bruxelles, du 30 octobre au 3 novembre 1911, sous les Auspices de M. E. Solvay*. Ed. by Paul Langevin and Maurice de Broglie. Paris: Gauthier-Villars, pp. 153–250.

Pohl, Robert W. and Peter Pringsheim (1914). *Die lichtelektrischen Erscheinungen*. Braunschweig: Vieweg.

Pringsheim, Peter (1922). "Über die Zerstörung der Fluoreszenzfähigkeit fluoreszierender Lösungen durch Licht und das photochemische Äquivalentgesetz". In: *Zeitschrift für Physik* 10, pp. 176–184.

Pringsheim, Peter (1923). "Über die photochemische Umwandlung fluoreszierender Farbstofflösungen". In: *Zeitschrift für Physik* 16, pp. 71–76.

Rabel, Gabriele (1919). "Farbenantagonismus oder die chemische und elektrische Polarität des Spektrums". In: *Zeitschrift für wissenschaftliche Photographie, Photophysik und Photochemie* 19, pp. 69–128.

Radel, Ernst (1920). "Ladungsbestimmungen an Nebelteilchen; ein Beitrag zur Frage der Existenz des elektrischen Elementarquantums". In: *Zeitschrift für Physik* 3, pp. 63–88.

Regener, Erich (1920). "Über die Ursache, welche bei den Versuchen von Hrn. F. Ehrenhaft die Existenz eines Subelektrons vortäuscht". In: *Sitzungsberichte der Preussischen Akademie der Wissenschaften*, pp. 632–641.

Richter, Steffen (1972). *Forschungsförderung in Deutschland 1920–1936. Dargestellt am Beispiel der Notgemeinschaft der Deutschen Wissenschaft und ihrem Wirken für das Fach Physik*. Düsseldorf: VDI-Verlag.

Rubens, Heinrich and Gerhard Michel (1921a). "Beitrag zur Prüfung der Planckschen Strahlungsformel". In: *Sitzungsberichte der Preussischen Akademie der Wissenschaften*, pp. 590–610.

Rubens, Heinrich and Gerhard Michel (1921b). "Prüfung der Planckschen Strahlungsformel". In: *Physikalische Zeitschrift* 22, pp. 569–577.

Schaefer, Clemens and Martha Schubert (1922). "Ultrarote Eigenfrequenzen der Silikate". In: *Zeitschrift für technische Physik* 3, pp. 201–204.

Schleiermacher, August and Richard Schachenmeier (1923) "Otto Lehmann". In: *Physikalische Zeitschrift* 24, pp. 289–291.

Schlüter, Steffen (1994). "Albert Einstein als Direktor des Kaiser-Wilhelm-Instituts für Physik". MA thesis. Berlin: Institut für Geschichtswissenschaft, Bereich Archivwissenschaft, Humboldt-Universität.

Schlüter, Steffen (1995). "Albert Einstein als Direktor des Kaiser-Wilhelm-Instituts in Berlin-Schöneberg". In: *Jahrbuch für Brandenburgische Landesgeschichte* 46, pp. 169–185.

Schröder, Wilhelm H. (1995). *Sozialdemokratische Parlamentarier in den Deutschen Reichs- und Landtagen 1867–1933*. Düsseldorf: Droste.

Seelig, Carl (1960). *Albert Einstein. Leben und Werk eines Genies unserer Zeit*. Zurich: Europa Verlag.

Seeliger, Rudolf (1920). "Über die Anregungsbedingungen der Quecksilberlinien". In: *Zeitschrift für Physik* 2, pp. 405–414.

Seeliger, Rudolf (1922a). "Anregung der Atome zur Lichtemission durch Elektronenstoß. V. Das Verhalten von Kombinationslinien". In: *Zeitschrift für Physik* 11, pp. 197–200.

Seeliger, Rudolf (1922b). "Über die Lichtemission der Glimmentladung". In: *Annalen der Physik* 67, pp. 352–358.

Seeliger, Rudolf and Georg Mierdel (1921). "Anregung der Atome zur Lichtemission durch Elektronenstoß II. Spektroskopische Studien an der Neon-Glimmlampe". In: *Zeitschrift für Physik* 5, pp. 182–187.

Seeliger, Rudolf and Dorothea Thaer (1921). "Die Bogen- und Funkenspektra der Alkalien, Erdalkalien und Erden". In: *Annalen der Physik* 65, pp. 423–448.

Seemann, Hugo (1921). "Ein Präzisions-Röntgenspektrograph". In: *Physikalische Zeitschrift* 22, pp. 580–581.

Starkulla, Heinz (1971). "Fleischer, Richard". In: *Neue Deutsche Biographie*. Vol. 5. Berlin: Duncker & Humblot, pp. 233–234.

Stern, Otto (1922). "Über den experimentellen Nachweis der räumlichen Quantelung im elektrischen Feld". In: *Physikalische Zeitschrift* 23, pp. 476–481.

Steubing, Walter (1919a). "Eine diamagnetische Erscheinung in leuchtendem Stickstoff und magnetisches Verhalten seiner Bandenspektra". In: *Physikalische Zeitschrift* 20, pp. 512–519.

Steubing, Walter (1919b). "Spektrale Intensitätsverschiebung und Schwächung der Jodfluoreszenz durch ein magnetisches Feld". In: *Annalen der Physik* 58, pp. 55–104.

Steubing, Walter (1921). "Temperatur und Bandenspektrum". In: *Physikalische Zeitschrift* 22, pp. 507–511.

Steubing, Walter (1922). "Die Spektra von Argon, Jod und Stickstoff im elektrischen Feld". In: *Physikalische Zeitschrift* 23, pp. 427–432.

Stintzing, Hugo (1938). "25 Jahre Röntgen-Strahlen-Beugung. Dem Andenken Paul Knippings, ihres Mitentdeckers und Gründers des Röntgeninstituts der Technischen Hochschule Darmstadt". In: *Zeitschrift für technische Physik* 19, pp. 104–105.

Stobbe, Hans (1936–1940). *J. C. Poggendorffs biographisch-literarisches Handwörterbuch*. Band VI: *1923 bis 1931*. Berlin: Verlag Chemie.

Suhrmann, Rudolf (1922). "Rote Grenze und spektrale Verteilung der Lichtelektrizität des Platins in ihrer Abhängigkeit vom Gasgehalt". In: *Annalen der Physik* 67, pp. 43–68.

Trautz, Max and Otto Großkinsky (1922). "Die Messung der spezifischen Wärme C_v von Gasen mittels der Differentialmethode. I. Mitteilung". In: *Annalen der Physik* 67, pp. 462–526.

Urbschat, Rudolf (1921). "Die reversible magnetische Permeabilität des Eisens bei hohen Frequenzen". In: *Zeitschrift für Physik* 7, pp. 260–267.

Valentiner, Siegfried (1929). "Ernst Wagner". In: *Physikalische Zeitschrift* 30, pp. 281–290.

vom Brocke, Bernhard (1990). "Die Kaiser-Wilhelm-Gesellschaft in der Weimarer Republik. Ausbau zu einer gesamtdeutschen Forschungsorganisation (1918–1933)". In: *Forschung im Spannungsfeld von Politik und Gesellschaft. Geschichte und Struktur der Kaiser-Wilhelm-/Max-Planck-Gesellschaft*. Ed. by Rudolf Vierhaus and Bernhard vom Brocke. Stuttgart: Deutsche Verlagsanstalt, pp. 197–355.

von Harnack, Adolf (1923). *Kaiser-Wilhelm-Gesellschaft zur Förderung der Wissenschaften. Bericht Oktober 1922–Dezember 1923*. Burg/Magdeburg: Hopfer.

von Harnack, Adolf ed. (1928). *Handbuch der Kaiser Wilhelm-Gesellschaft zur Förderung der Wissenschaften*. Berlin: Hobbing.

von Klüber, Harald (1966). "Grotrian, Walter". In: *Neue Deutsche Biographie*. Vol. 7. Berlin: Duncker & Humblot, pp. 169–170.

von Schweidler, Egon (1915). "Über die möglichen Quellen der Hessschen Strahlung". In: *Arbeiten aus den Gebieten der Physik, Mathematik, Chemie. Festschrift Julius Elster und Hans Geitel zum sechzigsten Geburtstag gewidmet von Freunden und Schülern*. Braunschweig: Vieweg, pp. 411–419.

Wagner, Ernst (1918). "Spektraluntersuchungen an Röntgenstrahlen. Über die Messung der Planckschen Quantenkonstante h aus dem zur Erzeugung homogener Bremsstrahlung notwendigen Minimumpotential". In: *Annalen der Physik* 57, pp. 401–470.

Wagner, Ernst and Helmuth Kulenkampff (1922). "Die Intensität von Röntgenstrahlen verschiedener Wellenlänge an Kalkspat und Steinsalz". In: *Annalen der Physik* 68, pp. 369–413.

Warburg, Emil, Gustav E. Leithäuser, et al. (1913). "Über die Konstante c des Wien-Planckschen Strahlungsgesetzes". In: *Annalen der Physik* 40, pp. 609–634.

Warburg, Emil and Carl Müller (1915). "Über die Konstante c des Wien-Planckschen Strahlungsgesetzes". In: *Annalen der Physik* 48, pp. 410–432.

Wazeck, Milena (2009). *Einsteins Gegner. Die öffentliche Kontroverse um die Relativitätstheorie in den 1920er Jahren*. Frankfurt am Main: Campus Verlag.

Weigert, Fritz (1919a). "Über einen neuen Effekt der Strahlung in lichtempfindlichen Schichten (Erste Mitteilung)". In: *Verhandlungen der Deutschen Physikalischen Gesellschaft* 21, pp. 479–491.

Weigert, Fritz (1919b). "Über einen neuen Effekt der Strahlung (Zweite Mitteilung)". In: *Verhandlungen der Deutschen Physikalischen Gesellschaft* 21, pp. 615–622.

Weigert, Fritz (1919c). "Über einen neuen Effekt der Strahlung (Dritte Mitteilung)". In: *Verhandlungen der Deutschen Physikalischen Gesellschaft* 21, pp. 623–631.

Weigert, Fritz (1921). "Über einen neuen Effekt der Strahlung (Vierte Mitteilung)". In: *Zeitschrift für Physik* 5, pp. 410–427.

Weigert, Fritz (1922). "Über Fluoreszenz, photochemische Wirkung und das Einsteinsche Gesetz". In: *Zeitschrift für Physik* 10, pp. 349–351.

Weigert, Fritz and Gerhard Käppler (1924). "Polarisierte Fluoreszenz in Farbstofflösungen". In: *Zeitschrift für Physik* 25, pp. 99–117.

Weinmeister, Paul (1925–1926). *J. C. Poggendorffs biographisch-literarisches Handwörterbuch.* Band V: *1904 bis 1922.* Leipzig: Verlag Chemie.

Wendt, Georg (1917). "Spektralanalytische Untersuchungen an Kanalstrahlen von Kohlenstoff, Silicium und Bor". In: *Annalen der Physik* 52, pp. 761–774.

Wendt, Georg and R. A. Wetzel (1916). "Beobachtungen über den Effekt des elektrischen Feldes auf die Tripletserien des Quecksilbers und die Dubletserien des Aluminiums". In: *Annalen der Physik* 50, pp. 419–432.

Westphal, Wilhelm (1919a). "Über das Radiometer". In: *Verhandlungen der Deutschen Physikalischen Gesellschaft* 21, pp. 129–143.

Westphal, Wilhelm (1919b). "Zur Theorie des Radiometers". In: *Verhandlungen der Deutschen Physikalischen Gesellschaft* 21, p. 669.

Westphal, Wilhelm (1920a). "Messungen am Radiometer". In: *Zeitschrift für Physik* 1, pp. 92–100.

Westphal, Wilhelm (1920b). "Messungen am Radiometer II". In: *Zeitschrift für Physik* 1, pp. 431–438.

Westphal, Wilhelm (1921). "Messungen am Radiometer III. Über ein Quarzfaden-Radiometer". In: *Zeitschrift für Physik* 4, pp. 221–225.

Westphal, Wilhelm and Walther Gerlach (1919). "Über positive und negative Radiometerwirkungen". In: *Verhandlungen der Deutschen Physikalischen Gesellschaft* 21, pp. 218–226.

Wigand, Albert (1924). "Die durchdringende Strahlung". In: *Physikalische Zeitschrift* 25, pp. 445–463.

Wilhelm, Johannes P. (1987). "Rudolf Seeliger und die Plasmaphysik". In: *Sitzungsberichte der Akademie der Wissenschaften der DDR. Mathematik-Naturwissenschaft-Technik* 3.N, pp. 1–32.

Witt, Peter-Christian (1990). "Wissenschaftsfinanzierung zwischen Inflation und Deflation: Die Kaiser-Wilhelm-Gesellschaft 1918/19 bis 1934/35". In: *Forschung im Spannungsfeld von Politik und Gesellschaft. Geschichte und Struktur der Kaiser-Wilhelm-/Max-Planck-Gesellschaft.* Ed. by Rudolf Vierhaus and Bernhard vom Brocke. Stuttgart: Deutsche Verlagsanstalt, pp. 579–656.

Wolter, Kurt (1921). "Über Ladungsbestimmungen an Nebelteilchen bei 1 bis 9 Atm. Gasdruck". In: *Zeitschrift für Physik* 6, pp. 339–351.

Zaunick, Rudolf and Hans Salié (1956–1962). *J. C. Poggendorffs biographisch-literarisches Handwörterbuch der exakten Naturwissenschaften.* Band VIIa: *Berichtsjahre 1932 bis 1953.* Berlin: Akademie-Verlag.

Chapter 4
Einstein's Directorship: An Evaluation

Abstract This final chapter evaluates the activities of the Kaiser-Wilhelm-Institut für physikalische Forschung and its role in the advancement of physics. It is concluded that the institute only partially fulfilled the expectations of its promoters because of the waning interest in quantum physics on the part of its director and board, and also because of Einstein's refusal to exert scientific leadership.

Keywords Einstein's research 1917–1922 · Science policy in Germany

4.1 The Institute's Spending Policy

First let us consider the institute's funding policy from a financial and quantitative point of view.[1] As a matter of fact, the KWI für Physik did not implement the originally intended policy of employing, in Nernst's words, "major funding"[2] for "experimental works on a larger scale,"[3] although the intention to operate in this way was never formally abandoned; on the contrary it was often restated. Since the beginning, the funds were quite conspicuous, certainly enough to finance a major research program,[4] but it was soon decided to follow a policy of saving in order to have more money for eventual larger research projects. In the spring of 1919—at the real start of the institute's activity after the war and concurrently with the decision to distribute funds with the sole aim of relaunching physics research—the Direktorium decided to spend only as much as would become available during the

---

[1]For an analysis, from a different point of view, of the support policy for physical research in Germany from 1920 on, see Forman (1974).

[2]Minutes of the meeting, 9 January 1914, GStA, I. HA, Rep. 76 76 V c, Sekt. 2, Tit. 23, Litt. A, Nr. 116, p. 18.

[3]"Experimentelle Arbeiten grösseren Stils" ("Im nachfolgenden beehre ich mich ..." 2 March 1914, SBB, Acta Preussische Staatsbibliothek, Generaldirektion, Kaiser-Wilhelm-Institute XXVI).

[4]Remember that, at the beginning, the annual income of 75,000 marks largely exceeded the incomes of the physics and chemistry institutes of the Berlin University, at which Rubens and Nernst carried out their systematic investigations (see Sect. 2.3 and Tables 2.1 and 2.2).

© The Author(s), under exclusive license to Springer Nature Switzerland AG 2020
H. Goenner and G. Castagnetti, *Establishing Quantum Physics in Berlin*,
SpringerBriefs in History of Science and Technology,
https://doi.org/10.1007/978-3-030-63122-2_4

ongoing fiscal year and not to use the surplus from the previous one.[5] For the entire period 1919–1922, the annual reports and other documents reiterated that funds had been set aside "in order to be able to cope with some single larger and more expensive task,"[6] "as a reserve for large-scale endeavors,"[7] the intention being "to make available the existing funds, possibly undivided, for important investigations."[8] This was probably the reason why Einstein put some applicants off with the excuse of deficient funds, although money was largely available.[9] It would also explain why it was so controversial to give Kohn as well as Kallmann and Knipping the extra funds they needed.[10] As a consequence of this policy, during the period under examination expenditures were always inferior to incomes and the institute had more money at the end of every fiscal year than at the beginning.[11] Therefore sufficient funds were always available to launch a major research program. It was not the lack of funds that hindered the KWI für Physik in supporting more scientific research, but rather its lack of scientific initiative and the inability of the Direktorium to fulfill its task of suggesting and organizing large-scale research. As the institute never started any consistent research program of its own, the money was dispersed among many small projects. The declarations in the annual reports were mere lip service.

That Einstein himself was far removed from thinking in terms of large-scale research can be seen in his attitude towards research on the subject that mattered to him most. Despite his keen interest in the experimental testing of the theory of general relativity, he did not consider that expensive instrumentation or even a whole new research group would be necessary.[12] In view of the lack of adequate instruments in German observatories, and taking advantage of the public interest provoked by the successful English solar eclipse expeditions of May 1919, in December 1919 a public

---

[5] Einstein, "Erläuterungen zu dem Haushaltsplan," undated [sent to von Siemens on 9 May 1919], AMPG, I. Abt., Rep. 34, Nr. 13.

[6] Report of the KWI für Physik 1 April 1920–31 March 1921, AMPG, I. Abt., Rep. 34, Nr. 13.

[7] Einstein to von Harnack, 13 July 1921, AMPG, I. Abt., Rep. 34, Nr. 13.

[8] "Das Kaiser-Wilhelm-Institut für Physik," 28 July 1922, AMPG, I. Abt., Rep. 34, Nr. 13; Kaiser-Wilhelm-Gesellschaft (1922, 28–29). See also Report of the KWI für Physik 1 April 1919–31 March 1920, AMPG, I. Abt., Rep. 1 A, Nr. 1665, p. 48; Kirsten and Treder (1979, 152, vol. 1).

[9] For example, due to a lack of money Einstein put Hallwachs and Jensen off in May 1919 and Born in September 1919 although the institute still had 65,000 marks to distribute at the end of September 1919 after disbursing all the grants approved at the Direktorium meeting of 24 April 1919 (Einstein to Hallwachs, 16 May 1919, AMPG, I. Abt., Rep. 34, Nr. 4, folder Hallwachs; Einstein to Jensen, 16 May 1919, AMPG, I. Abt., Rep. 34, Nr. 5, folder Jensen; Einstein to Hedwig Born, 1 September 1919, in Einstein and Born (1969, 30); bank statement 1 April–31 December 1919, AMPG, I. Abt., Rep. 34, Nr. 8, folder Mendelssohn).

[10] See page 79. The grants for Kohn and for Kallmann and Knipping were decided in July and September 1922 respectively. At the end of September 1922, the institute still had more than 300,000 marks to spend (Bank statement 1 April–30 September 1922, AMPG, I. Abt., Rep. 34, Nr. 8, folder Mendelssohn).

[11] See Table 3.1. For 1922/23 this is true only in nominal terms, because of inflation.

[12] In fact, Einstein "knew nothing of the astrophysical literature of the time" and was totally dependent upon the advice of friendly experts in order to understand the technical and theoretical problems relating to astrophysical observations (Hentschel 1998, 469).

appeal for contributions to an "Einstein-Spende" (Einstein Donation) was initiated by Freundlich and signed by Haber, Harnack, Müller, Nernst, Planck, Rubens, Struve, and Warburg.[13] The money was to be used for the construction of an observatory whose single explicit task was to collect empirical data to test Einstein's theory of gravitation. In a very few months, a considerable amount of money was gathered from private sources as well as from the Prussian Government (Hentschel 1992, 66–67). But Einstein, moved by altruistic feelings in the difficult economical situation at the time, had doubts about the need for large funds, as may be seen from his letter of thanks to the Prussian Minister of Education:

> Will such a decision [to allocate to the Einstein-Spende 150,000 marks from the state budget], in the present times of utter misery, not arouse bitter feelings among the public? I think that we could effectively support research in the field of general relativity even without additional state funds if only the astronomical observatories and astronomers of the country would employ part of their instruments and manpower to this end.[14]

Returning to the KWI für Physik, once the decision to follow a restrictive distribution policy had been taken, it seems that no further discussion in this respect ever took place. It may be too simplistic, but a hint in a letter from Warburg makes it plausible to suppose that the Direktorium members did not really know how much money they had at their disposal. Expressing his disapproval at having to pay more for the notorious apparatus needed by Kohn, Warburg argued that Kohn's research subject (photoelectric effect in gases) did not seem very interesting to him "as to expend almost the entire annual income of the institute (I think 70,000 marks)" for it.[15] Warburg did not know that the yearly endowments had been increased twice since the beginning, the last time in March 1922 to a total of 314,000 marks. It is doubtful whether Einstein reported regularly on the bank statements sent to him every three months.

On the other hand, it must be noted that, once a project had been approved, the applicant almost always received the amount he had requested. The cases of Trautz, Kohn, and Kallmann and Knipping, in which the grants did not match the request (Trautz) or were controversial, were exceptions. It was not the institute but the researchers themselves who determined, by the amount of their requests, the funding policy.

The great majority of funds remained in the State of Prussia: less than ten per cent of total expenditure went to researchers outside the state.[16] Within Prussia, Berlin and Göttingen got the biggest pieces of the pi.e. This was not the result of a policy preference, for which there is no documental evidence anyway, but rather a consequence of the fact that a little less than half of all German universities, among them some of the most important, were situated in Prussia (Minerva 1920; Forman 1968, 59a–59b). The universities of the other German states also had their own local

---

[13] "Albert Einstein-Spende," GStA, I. HA, Rep. 76 76 V c, Sect. 1, Tit. 11, Teil 5 c, Nr. 55, pp. 8–9; Kirsten and Treder (1979, 177, vol. 1).

[14] Einstein to Konrad Haenisch, 6 December 1919, in Einstein (2004, 274).

[15] Warburg to Einstein, 2 June 1922, AMPG, I. Abt., Rep. 34, Nr. 7, folder Kohn. See page 79.

[16] See Appendix.

sources of public and private support, so that their researchers had at first no reason to approach the KWI für Physik for funds. Feelings of local pride could also have prevented scientists from asking for help from an institution that even in its name gave testimony to the Prussian dynasty and hegemony.

4.2   The Scientists Funded

Most of the 48 scientists funded by the KWI für Physik during Einstein's directorship were ordinary or extraordinary professors (26, more than 50%) and *Privatdozenten* (12); fewer were assistants or had no academic position (10). Many of those belonging to junior group (*Privatdozenten* and assistants) later obtained decent academic positions within the German universities, so that roughly 80% of the total number of recipients were or became acknowledged members of the physics community. In contrast, Einstein's closest collaborators, Freundlich and Grommer, met with less success in German academia. Freundlich had to face the opposition of his colleagues; he became director of a department of the Astrophysikalisches Observatorium, dubbed the Einstein-Institut, but never obtained a chair in Germany (Hentschel 1992). Grommer had a precarious existence until his move to Minsk in 1928, where he became professor and member of the local Academy of Sciences (Pais 1982, 487–488). Bachem and Kohn, too, had the chance of an academic career only after emigration. Since all four of them were Jews, it is likely that anti-Semitism may also have thwarted their ambitions. Only Wendt left scientific research and is not recorded in the Poggendorffs bibliographies (Weinmeister 1925–1926; Stobbe 1936–1940; Zaunick and Salié 1956–1962).[17]

A major portion of the scientists supported by the KWI für Physik were already successful researchers, whose achievements were acknowledged by their contemporaries: Roughly 40% of the scientists supported are referred to in the eighth edition of Nernst's textbook on physical chemistry (Nernst 1921b), a third in Sommerfeld's seminal presentation of the state of the art in atomic physics (Sommerfeld 1922), 20% in Born's book concerning the application of Bohr's atomic model to solid state physics (Born 1923), and 20% in Pringsheim's much more specialized review of fluorescence (Pringsheim 1928).

As we showed, funds did not go to a preselected group of excellent scientists but rather to various people who, for different reasons, needed help and felt they could get it from the KWI für Physik. Some of them were involved directly or indirectly in research initiated by Direktorium members; some, like Franck and Hettner, certainly knew that their requests would be favorably received. But most of the applicants turned to the institute as one possible way to obtain funds without having any particular connection to its steering board. It is therefore not possible to speak about any set of supported scientists as a group forged by scientific connections. Nevertheless, it is

---

[17]Biographical and bibliographical information is taken from the Poggendorffs bibliographies (Weinmeister 1925–1926; Stobbe 1936–1940; Zaunick and Salié 1956–1962).

worth establishing, so to say post factum, possible relations among the scientists as a contribution to understanding the course of physics research at that time and the role played by the KWI für Physik. We will consider the "educational kinship" given by writing a doctoral thesis or by being an assistant under someone's supervision, and the "scientific relationships" determined by similar working methods, by the use of the same measuring procedures or instruments, or by very close scientific interests. Of course, most of the scientists will not present a "pure" affiliation but belong to more than one such network. Research projects similar to those funded by the KWI für Physik could have developed in all of these loosely defined scientific networks.

With regard to student-teacher relationships, let us first consider those relating to members of the Direktorium. Several scientists were former students of Warburg and Rubens. Warburg tutored Franck, Schaefer, and Pohl—who in turn was the *Doktorvater* (doctoral advisor) of Gudden. Rubens tutored Koenigsberger and Hettner, who was still his assistant. Westphal was also a former student of Rubens although he obtained his Ph.D. under Wehnelt's supervision, while Regener, who had been supervised by Drude, was a former student of Warburg. Günther and Krüger had been students of Nernst, with whom Günther still worked, and finally Kallmann had been tutored by Planck. As to constellations outside the Direktorium, some clusters can be detected. Many applicants came from Roentgen's school in Munich: Knipping, Koch, Magnus, Pringsheim, and Wagner. Bachem and Grebe had been students of Kayser in Bonn; Debye and Seeliger attained their degrees in Munich with Sommerfeld; Füchtbauer and Trautz with Ostwald in Leipzig; Gerlach and Wendt had been students of Paschen. Considering—without claim to completeness—the professional relations established after obtaining doctoral degrees, a group takes shape around Stark, with whom Kirschbaum, Steubing, and Wendt worked as assistants in Aachen. Baeyer, Franck, Pohl, Pringsheim, Regener, and Westphal had been assistants at the Physics Institut of Berlin University under the direction of Warburg or Rubens (Hoffmann 1984). Born had worked with Lummer in Breslau, and Stern with Einstein in Zurich. Before going to Potsdam, Grotrian worked as assistant in Göttingen with Born and Stern. Himstedt had been assistant to Warburg at the time when the latter was professor in Freiburg, and Koenigsberger in turn had been assistant there under Himstedt's direction some years later. Kallmann and Knipping worked with Franck at Haber's institute in Berlin. Before the war, Kolhörster had been assistant to Wigand in Halle, and Wigand, in turn, assistant to Hallwachs in Dresden. Schaefer had been assistant to Rubens in Berlin and *Privatdozent* in Breslau with Lummer. Seeliger worked for several years at the Physikalisch-Technische Reichsanstalt in Berlin under Warburg's direction. Weigert had for several years been *Privatdozent* for physical chemistry at Nernst's institute in Berlin. Not surprisingly, these networks show frequent intersections at the names of Röntgen, Rubens, Stark, and Warburg, who had actually founded their own schools and were among the major figures in physics at the beginning of the century.

Concerning the relations built upon common scientific interests and methods, it appears that some of the supported scientists were very versatile people, who happened to work in the same field, if at different times. Einstein, Freundlich, Grebe, and Bachem were brought together by their interest in testing the spectral redshift predicted by general relativity. Of course, Einstein, Haber, Nernst, Born, and Debye were thoroughly linked, not only through their work concerning specific heats of materials. Einstein shared a common interest in photochemistry with Pringsheim, Warburg, and Weigert. Franck, Nernst, Pringsheim, Warburg, and Weigert were all concerned with fluorescence. Gerlach, while bound closely to Stern and molecular beams, joined Born in an investigation of lattice theory and also used the Debye-Scherrer method of X-ray diffraction. Like Westphal and Einstein, he also worked on radiometer physics. Kirchner, Kirschbaum, Knipping, Seemann, and Wagner worked in X-ray research; Försterling, Franck, Füchtbauer, Hettner, Kohn, Schaefer, Seeliger, and Steubing worked in one or the other aspect of atomic and molecular spectroscopy. The photoelectric effect brought together Einstein, Gerlach, Koch, Pohl, Pringsheim, and Hallwachs. A close network of interrelations becomes apparent from the mutual references in published papers.[18] Two focal points in the sense of the greatest inter-action can be detected: Warburg and Nernst. Around Warburg are grouped Franck, Pohl, Pringsheim, Schaefer, and Weigert. Around Nernst is a circle formed by Gün-ther, Kolhörster, Magnus, Trautz, and Wigand. A sub-focus could be seen in the work of Franck who has links to Born, Gerlach, Seeliger, and Steubing. Rubens is directly connected with Warburg due to the precision measurements of black-body radiation. He influenced the scientific problems studied by Hettner and Schaefer. Försterling refers to the research interests of Rubens, Nernst, Debye, and Born among others. Almost everyone refers to Einstein in one publication or the other.

4.3   How Effective Was the Funding?

As mentioned, Einstein did not agree with the funding policy of the Direktorium and his words to Loeb can be considered testimony to the fact that the money was not always allocated according to the criteria of scientific excellence. Nevertheless, Ein-stein's implicit accusation that the money was wasted is an exaggeration stemming from his own bitterness. In view of the group of outstanding people forming the core of those supported by the institute, the funds were surely not wasted. This is con-firmed by the number of papers published. On the basis of references in documents, acknowledgments in articles, and the strict connection between proposed projects and content of publications, we calculate that about 75 papers resulted from work supported by the KWI für Physik during Einstein's period, to which a dozen papers by Grotrian and about 25 papers by Franck's group should be added. The results were impressive not only in their numbers: among the supported research projects were some of the most important of the time because of their consequences for the advancement of physical knowledge, like the Stern-Gerlach experiments. Of course, the KWI was not the only donor and not always the largest. The universities as well as

---

[18]We systematically searched *Annalen der Physik, Physikalische Zeitschrift, Sitzungsberichte der Preussischen Akademie der Wissenschaften, Zeitschrift für Physik,* and *Zeitschrift für Physikalische Chemie.*

public and private donors gave their contribution, so that it is not possible to establish anything like a cost/benefit ratio relating only to the institute's funds. Nevertheless, it is clear that the KWI für Physik contributed substantially to the progress of physics in Germany. Still, in some cases the financing did not have satisfactory effects. Sometimes, the research came to a conclusion only after many years, so that its results were obsolete in the face of newly arising questions. Sometimes the researchers, as in the cases of Fürth, Hettner, and Kohn, were never able to complete their ambitious projects. The difficulties caused by inflation had surely hampered their work, but possibly also a lack of experimental skill.

4.4 The Scientific Program

This analysis of the institute's activities leads us to conclude that the KWI für Physik under Einstein's directorship did not fulfill one of the specific tasks envisaged by its founding fathers. Remember that the institute had been created for the purpose of concentrating funds, efforts, and human resources into a few, well-defined clusters of new questions arising in physics (Heisenberg 1971, 46–47). From this perspective, the Direktorium ought to have been a kind of "think tank" in charge of discussing and formulating research projects to be carried out by others. In fact, instead of being the motor for a research program at the forefront of new physics it became a funding agency without priorities.

The projects supported from 1918 to 1922 covered a wide range of topics that can be categorized as following:

- gravitation (spectral redshift, light deflection, solar eclipse expeditions);
- kinetic theory and radiation (specific heats, thermodynamics, non-equilibrium thermodynamics in chemical reactions, radiometry, thermometry, ultramicroscopic particles);
- materials science (X-ray structure analysis, magnetic properties of metals and alloys, optical, elastical and thermal constants;
- nuclear physics (cosmic rays, radioactivity);
- physics of the atmosphere;
- physical chemistry (photochemistry);
- quantum theory (Planck's radiation law, directional quantization in a magnetic field);
- radiation and the quantum (cathode/anode rays, X-rays, photoeffect, emission by collision of electrons and atoms, thermal emission, fluorescence);
- spectroscopy (line/band spectra, absorption, dispersion, broadening of spectral lines, influence of electric and magnetic field).

Since support for research in these fields was not distributed according to a planned strategy, it appears post factum that the funding policy of the KWI für Physik was only defined by a sort of "exclusion principle": no liquid crystals, no meteorology, no astrometry, etc. Nevertheless, the bulk of the proposed projects concerned research on subjects relating to quantum physics, thus reflecting the spontaneous development of interests within the physics community: Of all the projects supported from 1918 until 1922, research on spectroscopic subjects represented roughly 40% and research on other topics related to the quantum nature of matter amounted roughly to another 36%. Therefore three quarters of the funded projects, in principle and to at least a limited extent, contributed to the development of quantum physics. The result was an increase in empirical data concerning the quantum structure of matter. From this point of view, the KWI für Physik at least partially accomplished its task in that its money contributed to the advancement of physical knowledge in the direction set by the initiators. But the fact remains that they themselves were not able to show the way and take the lead.

The Direktorium members had many opportunities to meet and work out a common research program. After all, they belonged to a small, close-knit community of scholars meeting several times a week to hear lectures and reports on the advancement of physics. Besides the Academy meetings every Thursday and those of the Deutsche Physikalische Gesellschaft every other Friday, every Wednesday they attended the "physics colloquium for advanced students" at the Berlin University, an open seminar on the latest experimental and theoretical developments under Rubens's and later Laue's chairmanship. During this colloquium Einstein and Nernst were lively protagonists in discussions extending from questions of quantum physics to thermodynamics, from relativity to astrophysics (Frank 1949, 192–193; Kallmann 1966, 490–492; Rompe 1979; Rompe 1980). Even though we have no records to support our conjecture, it is plausible to suppose that in this context ideas and suggestions for a research program to be implemented with the help of the KWI für Physik could easily have been taken under consideration.

4.5   The Scientific Programs of the Direktorium Members

In the first section of this study we pointed out that there were several points of contact regarding the scientific interests among the five initiators of the KWI für Physik and its first director. Of course, each Direktorium member had his own specialty and therefore his own specific research questions. The step that needed to be taken was to integrate these questions into a common program. Actually, this integration did not take place because the paths taken by the various Direktorium members diverged. Each had his own view on the kind of questions that needed to be addressed urgently in order to surmount the crisis affecting classical physics. Some members even abandoned their interest in this crisis altogether. The composition and, so to speak, democratic functioning of the Direktorium, where all members were peers, as well as Einstein's lack of personal leadership, did not allow the determination of the kind of unitary policy that would have been possible in a hierarchically structured institute.

We cannot give here an exhaustive characterization of the paths taken by the Direktorium members in pursuit of their scientific goals. Nevertheless, the following brief discussion may be enough to explain why a common research program was impossible. Einstein's interests alone will be presented in greater detail.

After Einstein, Haber was the youngest member of the Direktorium, having been born in 1868. His intended role in the Direktorium was probably that of expert organizer and link to the development of physical chemistry. Doubtless, Haber expected Einstein's theoretical help and suggestions for the advancement of his own discipline, but, immediately after the war, he "at first did not return to pure scientific research."[19] Instead he became occupied mainly with the reorganization of his institute—which during the war had been fully dedicated to research for the military—and with other administrative activities for the German national and Prussian governments. Later, from 1920 until 1926 Haber's main engagement was the search for a suitable method to obtain gold from sea water in order to contribute to the payment of Germany's war reparations. Nevertheless, even if he did not take the scientific lead in day-to-day work, Haber had gathered in his institute a group of excellent scientists, some of whom, like Franck and Hertz, had already made their own contributions to the advancement of quantum physics some years earlier. Haber was certainly not expected to supply a research program for the KWI für Physik and in the post-war years was probably, because of his other priorities, not even interested in the formulation of such a program. But when he realized that the Direktorium was simply distributing money without a rationale, Haber obviously took advantage of this for his own collaborators. As reported above, due to his recommendation, first Franck, and then later Kallmann and Knipping received financial help from the KWI für Physik for research done at the KWI für physikalische Chemie und Elektrochemi.e. Haber would probably have been able to make fruitful use of the entire budget of Einstein's institute, if only he had been allowed to dispose of it freely.

Nernst[20] was fifteen years older than Einstein and still very active in the years after the First World War. Since the enunciation of his heat theorem at the end of 1905, Nernst's major interest had been to test it. For this purpose, he mobilized his institute's entire staff for a decade, harvesting more than a hundred papers on experimental results. Nernst's experience and ideas about directing a research institute certainly stood behind the declaration that the new KWI für Physik should promote the "working out of important and urgent physics problems according to a plan."[21] After the publication of his monograph on the heat theorem in 1918, Nernst turned his mind to problems of photochemistry and chemical kinetics. In the context of his efforts to prove the validity of his heat theorem as a general law of nature, Nernst worked on the problem of gas degeneracy and encouraged young co-workers like Günther to do experimental research on the hypothesis he had formulated. As mentioned, Günther's work was supported by the KWI für Physik. For a short period after the First World

---

[19]Bonhoeffer (1953, 5). For detailed accounts on Haber's activities at this time, see the related chapters in Stoltzenberg (1994), Szöllösi-Janze (1998).

[20]These considerations regarding Nernst are based on Bartel and Huebener (2007), Bodenstein (1942), Hiebert (1978).

[21]Press announcement of the establishment of the KWI für Physik, *Vossische Zeitung*, 16 December 1917, Morgen-Ausgabe, Nr. 641, 2. Beilage "Finanz- und Handelsblatt," [p. 3].

War, Nernst took up the question of whether Planck's formula is a strictly valid law. As reported, Nernst and Wulf's calculations concerning this formula induced the KWI für Physik to support with one of its first grants a series of new measurements under the supervision of Rubens and Warburg. However, the true focus of Nernst's attention during the twenties became the so-called "heat death of the universe" as an unavoidable consequence of the increase of entropy in the system (Nernst 1921a). Nernst began looking for any possible source of new energy that would impede entropic degeneration, and therefore became more and more interested in the study of cosmic rays and other astrophysical phenomena. For this purpose, he recruited Kolhörster and other researchers to work on "altitude radiation," also providing them with support from the KWI für Physik. Throughout Nernst's scientific career there appears a pattern: he first identified theoretical problems (mostly in the application of physical principles to chemistry), then set out his goals and subsequently initiated a research program to attain them. He also knew how to employ the means and staff of an institute in order to pursue his program and was willing to do it. But the other Direktorium members were probably not convinced that Nernst's research program on cosmic rays was worth supporting continuously and with increasing funds, so they supported it only sporadically. In fact, the KWI für Physik remained as uncommitted with regard to Nernst's research interests as it was with regard to Einstein's.

Before Laue's appointment, only two of the six Direktorium members, Einstein and Planck, were theoretical physicists. At the start of the KWI für Physik in 1917, Planck[22] was already 59 years old, the second eldest after Warburg. His formation as a physicist had taken place during the zenith of what is called classical physics, in the second half of the nineteenth century. Although his discovery of the radiation law in 1900 had contributed to the upset of this very system, during the first decades of the new century Planck directed his efforts mainly to integrate the new developments of physics into the old theoretical structure. As was his custom, he pursued this task working alone and never set doctoral candidates or co-workers to work on these new questions. Of course, his hypotheses from this time were debated, opened the way for new insights, and were also taken up in experimental research. But after the First World War the new physics took a different direction and Planck stood on the sidelines as a critical spectator.

Rubens was 14 years older than Einstein and a recognized "old master of the art of experimenting and measuring."[23] His lifework until 1917 had been the exploration of the far infrared region in order to prove the validity of the electromagnetic wave theory of light. It was because of his precision measurements of the distribution of radiation energy that Planck arrived at his own radiation law. In the subsequent years until his death in 1922, Rubens applied his experimental ability to problems relevant to quantum physics, especially to the confirmation of Planck's radiation law. As reported, for these measurements he repeatedly received financial support from

---

[22]For these considerations regarding Planck, see Born (1948–1949), von Laue (1948), von Laue (1958).

[23]Franck and Pohl (1922a, 378); see also Kangro (1975).

the KWI für Physik. The guiding rationale of Rubens's experimental physics was, of course, not the mere collection of facts but consisted in "confronting a theoretically founded world view, testing it in the light of experience in order to confirm or to disprove it" (Franck and Pohl 1922b, 1030). On the other hand, in all likelihood nobody would expect from Rubens ideas or experiments suited to building a new theoretical system on the ruins of the old one.

Warburg,[24] born in 1846, was the eldest Direktorium member. For many years, he worked systematically on the testing of Einstein's law of photochemical equivalence. Concurrently, together with a group of co-workers, he also made experimental measurements concerning Planck's radiation law. However, by the time the KWI für Physik had started its activities, Warburg was no longer at the peak of his scientific productivity. Although still interested in theoretical problems of quantum physics, his contribution to the institute's direction could only come from his great experience as a "master of the art of experimenting" (Franck 1931, 993) (as he, too, was considered) and as organizer of a major scientific institute. As we know, in 1919 he could not take part in the program of new measurements of Planck's radiation formula supported by the KWI and instead continued to study gas discharges and other photochemical phenomena until his death in 1931.

4.6   Einstein's Research Program and the KWI Für Physik

Einstein's scientific work[25] during the period 1917—1922 was dominated by four great themes, the pursuit of which constituted what we can call his research program:

- the theoretical consolidation and the experimental testing of general relativity;
- the extension of general relativity towards a unified theory including gravitation and electromagnetism;
- the explanation of the quantum phenomena through overdetermination of partial differential equations, in order also to derive the quantum theory from a unified field theory;
- the resolution of the wave-particle duality of electromagnetic radiation.

The work on each of these particular topics was inspired by a more general motivation, rooted in Einstein's natural philosophy (Renn 2006, 130). As he himself put it, "the real goal of [his] research has always been the simplification and unification of the system of theoretical physics" (Dukas and Hoffmann 1979, 12).

At the beginning of his directorship Einstein was concerned almost exclusively with the theoretical foundation and the experimental verification of the theory of general relativity. To the first concern belongs the work on the energy conservation law. We have seen that the KWI für Physik in 1919 granted Grommer compensation

---

[24]For these considerations regarding Warburg, see Einstein (1922a), Grüneisen (1926), Franck (1931), Ramsauer (1913).

[25]For a comprehensive exposition of the development of Einstein's scientific work, see Pais (1982).

almost certainly for his contribution to the proof of the energy conservation law. To the second concern belong the measurement of light deflection and the detection of the redshift of spectral lines in the gravitational field of a celestial body. Besides the scholarship for Freundlich and the support for Grebe and Bachem, the KWI für Physik financed research on relativistic gravitation only in the case of Ludendorff's eclipse expedition in September 1923,[26] after the end of Einstein's directorship.

> In my spare time, I continually brood over the quantum problem from the point of view of relativity. I don't believe that the theory can dispense with the continuum. But I am not able to give tangible form to my pet idea of interpreting quantum structure through an overdetermination with differential equations.[27]

At about the same time, he complained that "the electric field still stands unrelated [to the gravitational field]" because "an overdetermination does not turn out."[28] One year later, he communicated to Sommerfeld that he had found "a kind of supplement for the foundations of general relativity which is akin to Weyl's"[29] and to Lorentz that he had "made an attempt at a generalization of the theory [of relativity]"—which he, however, regarded with skepticism.[30] Thus, Einstein embarked on his long journey towards a classical field theory of matter by pointing out the possibility that the constituent particles of matter may be built out of a combination of gravitational and electromagnetic fields.[31] The KWI für Physik supported Einstein's research in this field only with the grant given to Grommer in 1921.

Throughout 1921, Einstein thought intensively about the wave-particle duality in electromagnetism and devised more than one experiment to rule out one aspect or the other. One way was to try to disprove Maxwell's theory and thus the wave aspect. At first Einstein reasoned that in high-temperature radiation the field intensity predicted by the theory should produce a Stark effect, that is, a broadening of the spectral lines emitted by the atoms in the field, large enough to be perceptible.[32] Some months later he devised the canal rays experiment carried out by Geiger and Bothe, for which Einstein received from his institute only 579.85 marks while in the same period Born, Franck, and Pohl secured 100,000 marks to buy equipment for their Physics Institute at the University of Göttingen.

Besides these main focuses of scientific activity, there were other subjects that attracted Einstein's attention. For example, in the spring of 1919, he took an interest in radiometer physics, pointing out a flaw in Westphal's theory, as discussed earlier.

---

[26]Laue to Kuratorium, 17 May 1923, AMPG, I. Abt., Rep. 1 A, Nr. 1659, p. 29; Hentschel (1992, 134).

[27]Einstein to Born, 3 March 1920, in Einstein (2004, 459–460, doc. 337).

[28]Einstein to Ehrenfest, 7 April 1920, in Einstein (2004, 497–498, doc. 371).

[29]Einstein to Sommerfeld, 9 March 1921, in Einstein and Sommerfeld (1968, 78).

[30]Einstein to Lorentz, 30 June 1921, in Einstein (2009, 206–207, doc. 136).

[31]See Goenner (2004, sects. 4.3.2, 7.3, and 11).

[32]Einstein to Lorentz, 1 January 1921, in Einstein (2009, 22–24, doc. 3); Einstein to Sommerfeld, 4 January 1921, in Einstein and Sommerfeld (1968, 77–78); Einstein to Ehrenfest, 20 January 1921, in Einstein (2009, 46–47, doc. 24); Einstein to Born, 31 January 1921, in Einstein and Born (1969, 77–78).

Feeling the need to investigate the matter more closely, he suggested to his cousin, Edith Einstein, that she write a doctoral dissertation on the subject and gave her scientific advice. He finally wrote a paper on the subject himself.[33] In 1920, he returned to the kinetic theory of gases, one of his main areas of research since his dissertation, with a paper on the propagation of sound in partially dissociated gases (Einstein 1920). Between 1921 and 1922, he also thought about the phenomenon of superconductivity in metals. For this reason, he suggested to Ehrenfest an experiment to find out whether the current is carried by chains of molecules whose electrons continually interchange, and finally published a paper containing his reflections on the subject.[34] At the same time, he collaborated with Grommer to seek a way to test the existence of zero-point energy.[35]

This summary exposition of Einstein's scientific interests is by no means complete. Nevertheless, it shows toward how many different areas Einstein directed his attention during these years. All the more striking appears how seldom the KWI für Physik supported research initiated by Einstein, who, for his part, never tried to combine his theoretical work and his suggestions for experiments in a program for a team of researchers that could be supported by his institute.

Clearly, only in very few cases of research projects suggested by other scientists did Einstein feel a connection with his own interests. He therefore followed those projects with special attention and wrote about them in his letters. One such case is Hettner's investigation into the structure of infrared gas spectra. Einstein was highly interested in this research project, as it extended his work on quantization of molecular eigenvibration from solids to gases, and he mentioned it in a letter to Ehrenfest.[36] Another case, mentioned in the same letter, is Regener's work on the determination of the elementary electric charge. Although it had never been the subject of direct investigation by Einstein, the question often drew his attention[37] because a "point-like quantity of electricity" (Einstein 1905, 909; Einstein 1989, 294, doc. 23) played a role in his early verification of the theory of special relativity and because, for a time, Einstein supposed a connection between the "quantum of electricity" and the quantum of energy (Einstein 1909, 822; Einstein 1989, 577, doc. 60). He probably saw the unitary electric charge as evidence for the atomistic conception of matter and electricity. In 1922, Einstein took an active interest in the Stern-Gerlach experiment. In a paper co-authored with Ehrenfest he suggested two explanations for the experimental result and raised a problem:

---

[33] Edith Einstein to Albert Einstein, 29 April 1919, in Einstein (2004, 47, 48, doc. 31); Paul Epstein to Einstein, 2 May 1919, in Einstein (2004, 49, 50, doc. 32); Edith Einstein to Albert Einstein, 4 December 1921, in Einstein (2009, 366, doc. 310); Edith Einstein to Albert Einstein, 5 March 1922, AEA, 9–196; E. Einstein (1922), Einstein (1924).

[34] Einstein to Ehrenfest, 2 September 1921, in Einstein (2009, 270–271, doc. 225); Einstein to Ehrenfest, 21 January 1922, AEA, 10–011; Einstein (1922b).

[35] Einstein to Ehrenfest, 1 September 1921, in Einstein (2009, 264–265, doc. 219); Einstein to Ehrenfest, 20 February 1922, AEA, 10–023.

[36] Einstein to Ehrenfest, undated [6 June 1920], in Einstein (2006, 297–298, doc. 46). See also Einstein to Lorentz, 15 June 1920, Einstein (2006, 312–313, doc. 56).

[37] See, e.g.., Einstein to Ehrenhaft, 20 August 1918, in Einstein (1998, 861–862, doc. 605).

The most interesting thing at present is the experiment by Stern and Gerlach. The orientation of atoms without collisions cannot be explained by means of radiation, according to current reasoning; an orientation should, by rights, last more than a hundred years. I made a little calculation about it with Ehrenfest.[38]

And yet, although his research program and the KWI's support policy did not coincide, Einstein did not have a merely bureaucratic attitude towards the scientific aspects of the grant applications. On the contrary, he took his function as referee very seriously. Although we do not have records of discussions on the projects, the existing correspondence with Fürth, Rabel, and Trautz shows that Einstein always examined the experimental setups proposed by the applicants. He even took pains to examine the "organ for vision" devised by the obvious outsider Schweigler, as well as Kaiser's "secret" plan to gain electric energy from heat.

4.7   Einstein and Scientific Research as a Collective Enterprise

After having compared the institute's activities in the first five years with the initial intentions concerning the scale of the research projects and the scientific program to be pursued, let us finally consider the purpose of organizing research as a collective enterprise. Not only in the founding proposal of February 1914, but also in other related documents, we read that the institute "will use the largest part of the sums at its disposal for the support of work carried out in other scientific institutions."[39] As we argued, this manner of organizing research, involving scientists operating at the borderline between physics and physical chemistry, was probably the way in which the institute's promoters thought it would be possible to find a solution to the crisis produced in physics by the discovery of Planck's quantum.

At an early stage, Einstein was in principle well disposed towards the idea of team work: "The matter of an institute for me has been postponed until after my coming to Berlin. In fact, it would be good if I were to get some sort of institute; I could then work together with others instead of only by myself. This would be much more to my liking."[40] It seems, though, that he was not aware of the intrinsic need for the particular way of promoting and organizing research that was envisaged. The point was not only that one should work together with others, but also that the leader of the enterprise ought to formulate a working program (a conceptual task) and assemble people to work it out (an organizational problem). The story went differently.

The fact is that Einstein, because of his conception of knowledge production as being mainly due to individual creativity, did not really feel the need for collective scientific work and even less the need to plan work for a team. On many occasions

---

[38] Einstein to Born, undated [May 1922], in Einstein and Born (1969, 102); Einstein and Ehrenfest (1922).

[39] Prussian Ministry of Education to Prussian Ministry of Finance, 2 July 1914, GStA, I. HA, Rep. 76 76 V c, Sekt. 2, Tit. 23, Litt. A, Nr. 116, pp. 21–24.

[40] Einstein to Elsa Einstein, undated [7 November 1913], in Einstein (1993, 565, doc. 482).

Einstein asserted that, to his mind, the most important factor in science is the individual researcher with excellent ideas and skills, with creativity and intuition.[41] Of course, he was aware that, in view of the constant expansion and specialization of physical knowledge, there were problems that surpassed the capabilities of a single researcher. We know that throughout his scientific life Einstein worked with others on tackling specific questions as well as on writing papers.[42] It seems, though, that Einstein's collaborations almost always consisted in unequal relations between one person generating ideas and another elaborating on them, be it through calculational help or through experimental expertise: something very different from teamwork. Einstein's stress on individual creativity prevented him from developing an interest in structural questions connected with cooperative efforts in science. In talking to Rockefeller he once said "I put my faith in intuition" while the industrialist replied "I [...] on organization" (Nathan and Norden 1968, 157). Einstein left the organizational problems to others.

This was not without consequences for his attitude towards the institute and its tasks. In his answer to a circular from the President of the Notgemeinschaft inquiring about scientific projects to be supported, Einstein wrote: "I would however not dare to point to problems whose tackling seems to require support in the first place; this would be presumptuous. Nevertheless, I can give you names of researchers who, I think, promise important contributions."[43] He then listed Kossel, Franck, Stern, "Vollmer," and Gerlach: promising younger scholars, either experimental or theoretical physicists, working in very different fields.[44] For Einstein, an excellent physicist by definition brought with him excellent research projects. It is therefore understandable why he did not care to elaborate a research program for the KWI für Physik. It was not only that the scientific interests of the Direktorium members diverged considerably, making difficult the formulation of a program convincing to everyone. It was not only that Einstein was insensitive to the need for a collective effort. But, even more, he did not want to exert either scientific or managerial guidance. He was not willing to suggest what should be done. He considered it a "presumption." Years later, Einstein made a remark in which he presented himself as even less interested in the directorship of the KWI für Physik than we have assumed up to now. In commenting on a passage in the autobiography of his old friend János Plesch, he wrote:

> The story with the institute is another gracious lie. It is true, however, that I always knew how to manage it so that de facto I did not have the institute on my back. I just wanted to

---

[41] See, e.g., Einstein (1954, 13–14, 73–74, 77); Moszkowski (1921, 180–181). On Einstein's quasi-romantic conception of scientific activity, see also Einstein (1918).

[42] On Einstein's early scientific collaborations, see Pyenson (1985, 215–246).

[43] Draft of Einstein's letter to Schmidt-Ott, undated, written on the letter by Schmidt-Ott to Einstein, 19 July 1922, AMPG, I. Abt., Rep. 34, Nr. 9, folder Notgemeinschaft. Besides being Chairman of the Kuratorium of the KWI für Physik and of many other committees, Schmidt-Ott was also the President of the Notgemeinschaft since its establishment in October 1920 (Marsch 1994, 79).

[44] Walther Kossel had worked with Sommerfeld on X-rays and atomic structure (Möllenstedt 1980). Max Volmer, whose name Einstein misspelled, had worked with Stern on photochemistry (Blumtritt 1985).

have my head free, and I also did not want to have command over the activities of others
(nothing of the "Führer").[45]

Einstein's aversion to leadership is also confirmed by the fact that he made scant
use of his institutional power to engage scientists for research that mattered to him.
As we have seen, after the initial effort to assure a position for Freundlich and
aside from the minor financial support provided to Grommer, Grebe and Bachem,
Einstein did not use KWI für Physik's funds for people who worked for him. In
February 1922, although Grommer still had no academic position, Einstein did not
seize the offer of an assistantship extended to him by the Notgemeinschaft.[46] During
his time as director of the KWI für Physik, Einstein had no doctoral students and
no assistants at the university. There is a distinct discrepancy between Einstein's
public acclaim and the little actual influence he exerted. In general, we have no
hint of his ever having shaped the career of others with the aim of expanding his
personal sphere of influence, be it in terms of scientific ideas or of leadership in the
physics community. In contrast to other colleagues such as Nernst[47] in Berlin and
Sommerfeld[48] in Munich, he did not found a "school" either in the sense of a group
of people sharing his general, theoretical, and methodological approach nor in the
sense of former students strategically placed as professors in various universities. It
seems that Einstein never thought in terms of power; he was a man of inner values
and did not engage in the power plays typical of politics of any sort. Of course, he
was eager to promote his ideas and was pleased when other people took them up.
Indeed, the only influence he exerted, and barely another physicist in Germany at
that time had a comparable public status, was that of someone setting benchmarks
for excellence in the field. But he stood aside from the petty struggles for positions
and rank, being by disposition an "Einspänner," a loner (Pyenson 1985, 58–79).

4.8 Conclusion

The evidence is that the establishment of the KWI für Physik under Einstein's directorship had two major motivations: first of all, the awareness of a theoretical crisis
in physics and the wish to respond to it with a concerted effort; secondly, the wish
to bring to Berlin a great physicist who promised to contribute considerably to the
scientific effort as well as to add luster to the prestige of Berlin and Prussia. The
particular structure of the KWI für Physik was the result of a hybrid combination of

---

[45]Einstein to J. Plesch, 3 February 1944, displayed at the Stargardt auction, Berlin, 21–22 March
1996; see Stargardt (1996, 189–190, lot 443). The quoted sentence is not reported in the catalogue.
See also Plesch (1949, 135).

[46]Schmidt-Ott to Einstein, 28 February 1922, and draft of Einstein's letter to Schmidt-Ott, undated,
on the same page, AMPG, I. Abt., Rep. 34, Nr. 9, folder Notgemeinschaft.

[47]On the Nernst school, see Kant (1974).

[48]On the Sommerfeld school, see Eckert (1993), von Meyenn (1993), Seth (2010, 47–70), and
Eckert's contribution to this volume.

these endeavors with the agenda and idiosyncrasies of the individual actors. On the one hand, the aim of attacking a broad range of the conceptual problems presented by early quantum physics necessitated a cooperation between institutes and scientists from different disciplines. On the other hand, there were most probably doubts about Einstein's abilities to direct a research institute because of his professional specialization as a theoretician, without experience in experimental physics or in guiding collaborators. Finally, the promoters certainly wished to direct the institute's research program as well as its support policy in favor of their own collaborators. For all these reasons, the institute consisted at first only of a group of great scientists who were expected to formulate a common research program.

Meanwhile, at the time when the institute was established in October 1917 and even more so in 1919, when it effectively started its activities after the war, a shift in the scientific interests and research strategies of the institute's leaders had taken place. They were also losing touch with the advances in quantum physics. Above all, Einstein was neither willing nor able to exert scientific leadership in the expected sense, namely in a concerted attempt to solve the questions of quantum physics. Einstein's dedication to the fields of general relativity and, later, unified field theory, outweighed his interest and work in any other area of physics. He came more and more to rely solely on his own ideas and thus isolated himself from the physics community, losing technical competence. Even in fields close to his scientific interests like the testing of general relativity, he did not assume a leadership position. Moreover, because of the particular constitution of the steering board, in which six people with equal authority, equally great scientific self-esteem, and diverging perspectives had to share all decisions, a common research strategy could not be elaborated, and the institute became a funding agency without priorities. This change of function was in part also an emergency measure in the face of the poor economic situation in Germany after the lost war.

The proposals solicited by the KWI für Physik from the physics community were concerned with many different subfields of physics, notably with areas of current interest at that time such as spectroscopy, radiation, and the atomistic structure of matter. A few projects established a direct connection to the research interests of members of the Direktorium; others seemed to follow the paths long trodden by their proponents. If a leading, though not strictly applied criterion may be detected through the analysis of all the projects supported up to 1922/23, this was to exclude research outside the field of physics proper. Almost all the requests for reasonable purposes, even if controversial or of minor interest, were granted. This means that the funding policy of the KWI für Physik merely reacted to suggestions coming from ongoing physics research at the university institutes. In terms of science funding in general, however, the KWI für Physik did not fail: it even served as a model for the foundation of the Notgemeinschaft der Deutschen Wissenschaft. On the other hand, its role in this respect diminished after the establishment of the new funding agency.

At the first possible moment, that is, after gaining financial independence, Einstein withdrew himself from management of the institute. He had no interest in and very little influence on science politics relating to physics in Germany. Haber, Nernst, Planck, and Laue were the leaders in this field in Berlin; elsewhere Sommerfeld and others patiently wove the threads of their nets. Einstein seems to have been happy to be the ingenious theoretical physicist following the train of his creative ideas, and to pose as a public idol for the value of science.

4.9   Research Projects Supported by the KWI für Physik from October 1917 to March 1923

1 January 1918 – 31 December 1920

Scientists	Topic	Amount
E. Freundlich (Potsdam, Prussia)	Experimental and theoretical astronomical research to test the theory of general relativity and related questions	18,000 marks (6,000 marks annually for three years)

Fiscal year 1918/19

Scientists	Topic	Amount
P. Debye (Göttingen, Prussia)	Research by means of X-rays on the electron distribution inside the atom in crystals, in particular in diamond	16,030 marks
E. Freundlich (Potsdam, Prussia)	Testing the theory of general relativity	500 marks (reimbursement of expenses)

Fiscal year 1919/20

Scientists	Topic	Amount
E. Freundlich (Potsdam, Prussia)	Testing the theory of general relativity	2,850 marks (550 marks as reimbursement of expenses, 2,300 marks as extra pay)
K. Försterling (Danzig, Free City; from 1921 Jena, Saxony-Weimar)	Dependency upon temperature of the indices of refraction and absorption of metals in the infrared (project later abandoned); Zeeman/Paschen-Back effect in the hydrogen spectrum	2,000 marks
L. Grebe, A. Bachem (Bonn, Prussia)	Spectroscopic measurements on the cyanogen bands of the solar spectrum to verify the gravitational redshift	2,000 marks
J. Grommer (Berlin, Prussia)	Mathematical proof of the energy conservation law in the theory of general relativity	1,200 marks
P. Günther (Berlin, Prussia)	Viscosity of hydrogen and helium at low temperature; temperature dependence of the coefficient of viscosity	2,000 marks
W. Hammer, F. Himstedt (Freiburg, Baden)	Application of a new method for measuring capacities and constants of dielectricity in solving several problems; among others: testing of the Clausius-Mosotti relation in gases and liquids at different pressures; testing of a hypothesis of Debye concerning pre-existing dipoles	5,000 marks
W. Kaufmann (Königsberg, Prussia)	Measurements of the vacuum in transmitting tubes through ionization; magnetic measurements in high-frequency fields; measurements of constants of dielectricity and absorption at high frequencies	3,000 marks
F. Krüger, H. Bohlin (Danzig, Free City)	Research with a new X-ray technique on the crystallographic structure of metals and metal alloys	2,000 marks allocated but not disbursed
A. Magnus (Tübingen, Württemberg)	Specific heat coefficients of solids for high temperatures	3,000 marks
R. Pohl (Göttingen, Prussia)	Photoelectric effect and X-rays; influence of light on the electrical conductivity of crystals	5,000 marks
E. Regener (Berlin, Prussia)	Determination of the elementary electric charge	7,000 marks
H. Rubens, E. Warburg (Berlin, Prussia) (Warburg later renounced)	Verification of Planck's radiation law	Appropriated jointly: 20,000 marks; disbursed to Rubens during this period: 5,000 marks
R. Seeliger (Greifswald, Prussia)	Emission of light by atoms in gases; excitation conditions and critical excitation potential of selected lines	1,000 marks
H. Seemann (Würzburg, Bavaria)	Measurements of the polarization of X-rays to test Sommerfeld's theory of "Bremsstrahlung," i.e., continuous radiation emitted by decelerated electrons	3,000 marks
W. Steubing, G. Wendt (Aachen, Prussia). (In October 1919 Wendt was replaced by H. Kirschbaum)	Influence of a magnetic field on electron oscillations (Steubing) Influence of an electric field on the spectral lines of mercury and aluminium; research on the canal rays of different substances to advance the atomic theory and the knowledge of the light emission mechanism (Wendt) Influence of an electric field on nitrogen spectral lines (Kirschbaum and Steubing)	Allocated jointly 10,000 marks; disbursed during this period: 2,550 marks
E. Wagner (Munich, Bavaria)	Research on X-ray spectra to obtain a more precise determination of Planck's constant h	10,000 marks
W. Westphal (Berlin, Prussia)	Radiometric forces at high pressure; Knudsen's accommodation coefficient	3,000 marks
Physikalische Berichte	Running expenses	5,000 marks

Fiscal year 1920/21

Scientists	Topic	Amount
O. von Baeyer (Berlin, Prussia)	Determination of the elementary electric charge of droplets under different high pressures; determination of mobility and charge with the Einstein-Weiss method	5,000 marks
J. Franck, P. Knipping (Berlin, Prussia)	Collision of electrons with gas atoms	12,000 marks
C. Füchtbauer (Tübingen, Württemberg)	Intensity of spectral lines; dependency of the frequency of electronic transitions upon line width; causes of line broadening	7,000 marks
W. Hallwachs (Dresden, Saxony)	Influence of gas on the photoelectric effect	1,000 marks
G. Hettner (Berlin, Prussia)	Influence of an electric field on the absorption bands of gases in the infrared; testing of Planck's second quantum hypothesis	10,000 marks
P. P. Koch (Hamburg, Free City)	Intensity distribution in spectral lines; light effects on silver bromide	12,800 marks
H. Kohn, O. Lummer (Breslau, Prussia)	Photoelectric effect in gases and vapors; testing of Bohr's atomic model (Kohn) Zeeman effect, dispersion of light by gas atoms in the ultraviolet region (Lummer)	7,000 marks
R. Pohl, B. Gudden (Göttingen, Prussia)	Photoelectric effect in crystals; dependency of the constant of dielectricity upon irradiation by light; eigenvibrations of electrons in solids	8,000 marks
H. Rubens (Berlin, Prussia)	Verification of Planck's radiation law	1,000 marks
R. Seeliger (Greifswald, Prussia)	Continuation of research on light emission through collision of electrons with atoms in dissociated gases	2,000 marks
W. Steubing, H. Kirschbaum (Aachen, Prussia)	Influence of an electric field on nitrogen spectral lines; influence of temperature on the iodine spectral band	6,300 marks (disbursed from the sum allocated in the previous budget year)
E. Wagner (Munich, Bavaria)	Spectroscopic research on X-rays	1,500 marks
F. Weigert (Leipzig, Saxony)	Interaction of linearly-polarized light with light-sensitive layers of dye colloids	5,000 marks

Fiscal year 1921/22

Scientists	Topic	Amount
M. Born, J. Franck, R. Pohl (Göttingen, Prussia)	Collision of electrons and fluorescence (Franck) Photoelectric conductivity (Pohl) Dynamics of the crystal lattice (Born)	100,000 marks
A. Einstein (Berlin, Prussia)	Experiment with light emitted by canal rays to decide between the wave theory and the light-quantum theory of radiation	579.85 marks
R. Fürth (Prague, Czechoslovakia)	Measurement of the elementary electric charge with the Eötvös torsion balance	2,000 marks
W. Gerlach, O. Stern (Frankfurt, Prussia)	Influence of a magnetic field on band spectra of metal vapors	10,000 marks
J. Grommer (Berlin, Prussia)	Theoretical research in the field of relativity theory, probably for the extension of general relativity towards a unified field theory	2,000 marks
H. Rubens (Berlin, Prussia)	Verification of Planck's radiation law	1,000 marks
C. Schaefer (Marburg, Prussia)	Research on the infrared eigenvibrations of crystalline silicates to draw conclusions concerning the chemical structure of the material	10,000 marks
W. Steubing, H. Kirschbaum (Aachen, Prussia)	Spectroscopic research	850 marks (from the sum allocated in the budget year 1919/20)
W. Steubing (Aachen, Prussia)	Stark effect in band spectra	10,000 marks
M. Trautz (Heidelberg, Baden)	Measurements of the specific heat of many gases and for large temperature ranges with a new method	3,000 marks
E. Wagner (Munich, Bavaria)	Influence of potential and anticathode material on the emission of X-rays	2,500 marks
A. Wigand (Halle, Prussia)	Atmospheric electricity and radioactivity	2,000 marks
Physikalische Berichte	Printing costs	5,000 marks

Fiscal year 1922/23

Scientists	Topic	Amount
Deutsches Entomologisches Museum (Berlin, Prussia)		366,000 marks
W. Grotrian (Potsdam, Prussia)	Spectroscopic equipment of the Astrophysikalisches Observatorium Astrophysical research with new spectroscopic methods	400,000 marks
H. Kallmann, P. Knipping (Berlin, Prussia)	Ionization by electron bombardment	20,000 marks
J. Koenigsberger (Freiburg, Baden)	Scattering of hydrogen canal rays by gas atoms; absorption of canal rays by gases	4,000 marks
H. Kohn (Breslau, Prussia)	Photoelectric effect in gases and vapors	43,000 marks
W. Kolhörster (Berlin, Prussia)	Penetrating rays, i.e., cosmic radiation	110,000 marks
R. Pohl, B. Gudden (Göttingen, Prussia)	Photoelectric conductivity	90,839.20 marks
P. Pringsheim (Berlin, Prussia)	Absorption of light in mercury vapor	6,530 marks
C. Schaefer (Marburg, Prussia)	Infrared eigenvibrations of crystalline silicates	7,500 marks
R. Seeliger (Greifswald, Prussia)	Glow discharges in gases	1,500 marks
O. Stern (Rostock, Mecklenburg)	Magnetic properties of silver atoms; magnetic and electric deflection of molecular beams	20,000 marks
W. Steubing (Aachen, Prussia)	Stark effect in band spectra	2,500 marks

References

Bartel, Hans-Georg and Rudolf P. Huebener (2007). *Walther Nernst. Pioneer of Physics and Chemistry*. Singapore: World Scientific.

Blumtritt, Oskar (1985). *Max Volmer 1885–1965. Eine Biographie*. Berlin: Technische Universität Berlin.

Bodenstein, Max (1942). "Walther Nernst, 25.6.1864–18.11.1941". In: *Berichte der Deutschen Chemischen Gesellschaft, Abteilung A* 75, pp. 79–104.

Bonhoeffer, Karl-Friedrich (1953). "Fritz Habers wissenschaftliches Werk". In: *Zeitschrift für Elektrochemie* 57, pp. 2–6.

Born, Max (1923) *Atomtheorie des festen Zustandes (Dynamik der Kristallgitter)*. Leipzig: Teubner.

Born, Max (1948–1949). "Max Karl Ernst Ludwig Planck". In: *Obituary Notices of Fellows of the Royal Society* 6, pp. 161–188.

Dukas, Helen and Hoffmann Banesh, eds. (1979). *Albert Einstein. The Human Side*. Princeton: Princeton University Press.

Eckert, Michael (1993). *Die Atomphysiker: Eine Geschichte der theoretischen Physik am Beispiel der Sommerfeldschule*. Braunschweig: Vieweg.

Einstein, Albert (1905). "Zur Elektrodynamik bewegter Körper". In: *Annalen der Physik* 17, pp. 891–921. Reprinted in (Einstein 1989, doc. 23).

Einstein, Albert (1909). "Über die Entwicklung unserer Anschauungen über das Wesen und die Konstitution der Strahlung". In: *Physikalische Zeitschrift* 10, pp. 817–825. Reprinted in (Einstein 1989, doc. 60).

Einstein, Albert (1918). "Motive des Forschens". In: *Zu Max Plancks sechzigstem Geburtstag*. Ed. by Emil Warburg et al. Karlsruhe: Müllersche Hofbuchhandlung, pp. 29–32. Reprinted in (Einstein 2002, doc. 7).

Einstein, Albert (1920). "Schallausbreitung in teilweise dissoziierten Gasen". In: *Sitzungsberichte der Preussischen Akademie der Wissenschaften*, pp. 380–385. Reprinted in (Einstein 2002, doc. 39).

Einstein, Albert (1922a). "Emil Warburg als Forscher". In: *Die Naturwissenschaften* 10, pp. 823–828.

Einstein, Albert (1922b). "Theoretische Bemerkungen zur Supraleitung der Metalle". In: *Het Natuurkundig Laboratorium der Rijksuniversiteit te Leiden in de jaren 1904–1922: Gedenkboek aangeboden aan H. Kamerlingh Onnes*. Leiden: Ijdo, pp. 429–435.

Einstein, Albert (1924) "Zur Theorie der Radiometerkräfte". In: *Zeitschrift für Physik* 27, pp. 1–6.

Einstein, Albert (1954). *Ideas and Opinions*. New York: Crown Publishers.

Einstein, Albert (1989). *The Collected Papers*. Vol. 2: *The Swiss Years: Writings, 1900–1909*. Ed. by John Stachel. Princeton: Princeton University Press.

Einstein, Albert (1993). *The Collected Papers*. Vol. 5: *The Swiss Years: Correspondence, 1902–1914*. Ed. by Martin J. Klein, Anne J. Kox, and Robert Schulmann. Princeton: Princeton University Press.

Einstein, Albert (1998). *The Collected Papers*. Vol. 8: *The Berlin Years: Correspondence, 1914–1918*. Princeton: Princeton University Press.

Einstein, Albert (2002). *The Collected Papers*. Vol. 7: *The Berlin Years: Writings, 1918–1921*. Ed. by Michel Janssen et al. Princeton: Princeton University Press.

Einstein, Albert (2004). *The Collected Papers*. Vol. 9: *The Berlin Years: Correspondence, January 1919–April 1920*. Princeton: Princeton University Press.

Einstein, Albert (2006). *The Collected Papers*. Vol. 10: *The Berlin Years: Correspondence, May–December 1920 and Supplementary Correspondence, 1909–1920*. Ed. by Diana Kormos Buchhwald et al. Princeton: Princeton University Press.

Einstein, Albert (2009). *The Collected Papers*. Vol. 12: *The Berlin Years: Correspondence, January–December 1921*. Ed. by Diana Kormos Buchhwald et al. Princeton: Princeton University Press.

Einstein, Albert and Max Born (1969). *Briefwechsel 1916–1955. Kommentiert von Max Born*. Munich: Nymphenburger.

Einstein, Albert and Paul Ehrenfest (1922). "Quantentheoretische Bemerkungen zum Experiment von Stern und Gerlach". In: Zeitschrift für Physik 11, pp. 31–34.

Einstein, Albert and Arnold Sommerfeld (1968). *Briefwechsel. Sechzig Briefe aus dem goldenen Zeitalter der modernen Physik*. Ed. by Armin Hermann. Basel: Schwabe. Herausgegeben und kommentiert von Armin Hermann.

Einstein, Edith (1922). "Zur Theorie des Radiometers". In: *Annalen der Physik* 69, pp. 241–254.

Forman, Paul (1968). *The Environment and Practice of Atomic Physics in Weimar Germany*. Ann Arbor (Michigan): UMI.

Forman, Paul (1974). "The Financial Support and Political Alignement of Physicists in Weimar Germany". In: *Minerva* 12, pp. 39–66.

Franck, James (1931). "Emil Warburg zum Gedächtnis". In: *Die Naturwissenschaften* 19, pp. 993–997.

Franck, James and Robert Pohl (1922a). "Heinrich Rubens †". In: *Physikalische Zeitschrift* 23, pp. 377–382.

Franck, James and Robert Pohl (1922b). "Rubens und die Quantentheorie". In: *Die Naturwissenschaften* 10, pp. 1030–1033.

Frank, Philipp (1949). *Einstein. Sein Leben und seine Zeit*. Munich: List.

Goenner, Hubert (2004). "On the History of Unified Field Theories". In: *Living Reviews in Relativity* 7.2. https://link.springer.com/article/10.12942/lrr-2004-2 (accessed 12/21/2020).

Grüneisen, Eduard (1926). "Emil Warburg zum achtzigsten Geburtstage". In: *Die Naturwissenschaften* 14, pp. 203–207.

Heisenberg, Werner (1971). "Das Kaiser-Wilhelm-Institut für Physik. Geschichte eines Instituts". In: *Jahrbuch der Max-Planck-Gesellschaft zur Förderung der Wissenschaften*, pp. 46–89.

Hentschel, Klaus (1992). *Der Einstein-Turm: Erwin F. Freundlich und die Relativitätstheorie*. Heidelberg: Spektrum Akademischer Verlag.

Hentschel, Klaus (1998). *Zum Zusammenspiel von Instrument, Experiment und Theorie. Rotverschiebung im Sonnenspektrum und verwandte spektrale Verschiebungseffekte von 1880 bis 1960*. Hamburg: Kovac.

Hiebert, Erwin N. (1978). "Nernst, Hermann Walther". In: *Dictionary of Scientific Biography*. Vol. 15, Supplement I. Ed. by Charles C. Gillispie. New York: Scribner, pp. 432–453.

Hoffmann, Dieter (1984). "Die Physik an der Berliner Universität in der ersten Hälfte unseres Jahrhunderts. Zur personellen und institutionellen Entwicklung sowie wichtige Beziehungen mit anderen Institutionen physikalischer Forschung in Berlin". In: *Berliner Wissenschaftshistorische Kolloquien VIII: Die Entwicklung der Physik in Berlin*. Berlin: Akademie der Wissenschaften der DDR. Institut für Theorie, Geschichte und Organisation der Wissenschaft, pp. 5–29.

Kaiser-Wilhelm-Gesellschaft (1922). *Kaiser-Wilhelm-Gesellschaft zur Förderung der Wissenschaften. Jahresbericht April 1921–Oktober 1922*. Burg/Magdeburg: Hopfer.

Kallmann, Hartmut (1966). "Von den Anfängen der Quantentheorie. Eine persönliche Rückschau". In: *Physikalische Blätter* 22, pp. 489–500.

Kangro, Hans (1975). "Rubens, Heinrich". In: *Dictionary of Scientific Biography*. Vol. 11. Ed. by Charles C. Gillispie. New York: Scribner, pp. 581–585.

Kant, Horst (1974). "Zum Problem der Forschungsprofilierung am Beispiel der Nernstschen Schule während ihrer Berliner Zeit von 1905 bis 1914". In: *NTM Schriftenreihe für Geschichte der Naturwissenschaften, Technik und Medizin* 11, pp. 58–68.

Kirsten, Christa and Hans-Jürgen Treder, eds. (1979). *Albert Einstein in Berlin 1913–1933*. Berlin: Akademie-Verlag.

Marsch, Ulrich (1994). *Notgemeinschaft der DeutschenWissenschaft. Gründung und frühe Geschichte 1920–1925*. Frankfurt am Main: Lang.

Minerva. Jahrbuch der gelehrten Welt (1920). Vol. 24. Berlin: de Gruyter.

Möllenstedt, Gottfried (1980). "Kossel, Walther". In: *Neue Deutsche Biographie*. Vol 12. Berlin: Duncker & Humblot, pp. 616–617.

Moszkowski, Alexander (1921). *Einstein. Einblicke in seine Gedankenwelt*. Hamburg/Berlin: Hoffmann und Campe/Fontane.

Nathan, Otto and Heinz Norden (1968). *Einstein on Peace*. New York: Schoken.

Nernst, Walther (1921a). *Das Weltgebäude im Lichte der neueren Forschung*. Berlin: Springer.

Nernst, Walther (1921b). *Theoretische Chemie vom Standpunkte der Avogadroschen Regel und der Thermodynamik*. 8th–10th ed. Stuttgart: Enke.

Pais, Abraham (1982). *"Subtle is the Lord . . . " The Science and the Life of Albert Einstein*. Oxford: Oxford University Press.

Plesch, János (1949). *János. Ein Arzt erzählt sein Leben*. Munich: List.

Pringsheim, Peter (1928). *Fluoreszenz und Phosphoreszenz im Lichte der neueren Atomtheorie*. 3rd ed. Berlin: Springer.

Pyenson, Lewis (1985). *The Young Einstein. The Advent of Relativity*. Bristol: Hilger.

Ramsauer, Carl (1913). "Das physikalisch-radiologische Institut der Universität Heidelberg". In: *Frankurter Zeitung* 145, 25 July 1913. Reprinted in (Auer 1984, 57–62).

Renn, Jürgen (2006). *Auf den Schultern von Riesen und Zwergen. Einsteins unvollendete Revolution.* Weinheim: Wiley-VCH.

Rompe, Robert (1979). " '… eine faszinierende Persönlichkei' ". In: *Wissenschaft und Fortschritt* 29, pp. 46–48.

Rompe, Robert (1980). "Max von Laue und das Berliner physikalische Kolloquium". In: *Berliner Wissenschaftshistorische Kolloquien I. Über das persönliche und wissenschaftliche Wirken von Albert Einstein und Max von Laue.* Berlin: Akademie der Wissenschaften der DDR. Institut für Theorie, Geschichte und Organisation der Wissenschaft, pp. 111–117.

Seth, Suman (2010). *Crafting the Quantum. Arnold Sommerfeld and the Practice of Theory, 1890–1926.* Cambridge, MA: The MIT Press.

Sommerfeld, Arnold (1922). *Atombau und Spektrallinien.* 3rd ed. Braunschweig: Vieweg.

Stargardt, J. A. (1996). *Autographen aus allen Gebieten. Auktion am 21. und 22. März 1996.* Katalog 663. Berlin.

Stobbe, Hans (1936–1940). *J. C. Poggendorffs biographisch-literarisches Handwörterbuch.* Band VI: *1923 bis 1931.* Berlin: Verlag Chemie.

Stoltzenberg, Dietrich (1994). *Fritz Haber: Chemiker, Nobelpreisträger, Deutscher, Jude, eine Biographie.* Weinheim: VCH.

Szöllösi-Janze, Margit (1998). *Fritz Haber 1868–1934. Eine Biographie.* Munich: Beck.

von Laue, Max (1948). "Max Planck". In: *Die Naturwissenschaften* 35, pp. 1–7.

von Laue, Max (1958). "Zu Max Plancks 100. Geburtstage". In: *Die Naturwissenschaften* 45, pp. 221–226.

von Meyenn, Karl (1993). "Sommerfeld als Begründer einer Schule der Theoretischen Physik". In: *Naturwissenschaft und Technik in der Geschichte. 25 Jahre Lehrstuhl für Geschichte der Naturwissenschaft und Technik am Historischen Institut der Universität Stuttgart.* Ed. by Helmuth Albrecht. Stuttgart: Verlag für Geschichte der Naturwissenschaft und der Technik, pp. 241–261.

Weinmeister, Paul (1925–1926). *J. C. Poggendorffs biographisch-literarisches Handwörterbuch.* Band V: *1904 bis 1922.* Leipzig: Verlag Chemie.

Zaunick, Rudolf and Salié Hans (1956–1962). *J. C. Poggendorffs biographisch-literarisches Handwörterbuch der exakten Naturwissenschaften.* Band VIIa: *Berichtsjahre 1932 bis1953.* Berlin: Akademie-Verlag.

Appendix

Abbreviations

AAdW	Archiv der Berlin-Brandenburgischen Akademie der Wissenschaften, Berlin
AEA	Albert Einstein Archives, The Jewish National and University Library, Jerusalem
AMPG	Archiv zur Geschichte der Max-Planck-Gesellschaft, Berlin
GStA	Geheimes Staatsarchiv, Berlin
ETH	Eidgenossische Technische Hochschule, Zurich
SBB	Staatsbibliothek Berlin
KWG	Kaiser-Wilhelm-Gesellschaft
KWI	Kaiser-Wilhelm-Institut

Archival Sources

Archiv zur Geschichte der Max-Planck-Gesellschaft, Berlin	The documents of the Kaiser-Wilhelm-Gesellschaft relating to the Kaiser-Wilhelm-Institut für Physik are in I. Abt., Rep. 1 A, Nr. 1649–1671; the documents of the Kaiser-Wilhelm-Institut für Physik itself are in I. Abt., Rep. 34
Geheimes Staatsarchiv, Berlin	The documents relating to the Kaiser-Wilhelm-Gesellschaft and the Kaiser-Wilhelm-Institut für Physik are for the great part in the files of the Preussisches Kultusministerium (I. HA, Rep. 76), the Geheimes Zivilkabinett (I. HA, Rep. 89), and of the Nachlass Schmidt-Ott (I. HA, Rep. 92 Schmidt-Ott)

© The Author(s), under exclusive license to Springer Nature Switzerland AG 2020
H. Goenner and G. Castagnetti, *Establishing Quantum Physics in Berlin*,
SpringerBriefs in History of Science and Technology,
https://doi.org/10.1007/978-3-030-63122-2

Acknowledgments

We are grateful to Dieter Hoffmann, Michel Janssen, Horst Kant, Christoph Lehner, Jürgen Renn, Matthias Schemmel, and Skuli Sigurdsson for their suggestions and criticism, of which we took great advantage. Particular thanks go to Lindy Divarci and Jeremiah James for their patient work in correcting our English. For their help in archival research we would like to thank the staff of the Archiv zur Geschichte der Max-Planck-Gesellschaft, Berlin, and the staff of the Albert Einstein Archives at the Jewish National and University Library, Jerusalem. This study was made possible by a grant from the Berlin Senate and by the hospitality of the Max Planck Institute for the History of Science.

Uncited References

Albrecht, Helmuth, ed. (1993). *Naturwissenschaften und Technik in der Geschichte. 25 Jahre Lehrstuhl für Geschichte der Naturwissenschaft und Technik am Historischen Institut der Universität Stuttgart*. Stuttgart: Verlag für Geschichte der Naturwissenschaften und der Technik.

Doebler, Michael et al., eds. (1930). *Das akademische Deutschland*. Band III: *Die deutschen Hochschulen in ihren Beziehungen zur Gegenwartskultur*. Berlin: Weller.

Einstein, Albert (1996). *The Collected Papers*. Vol. 6: *The Berlin Years: Writings, 1914–1917*. Ed. by Anne J. Kox, Martin J. Klein, and Robert Schulmann. Princeton: Princeton University Press.

Einstein, Albert (2002). *The Collected Papers*. Vol. 7: *The Berlin Years: Writings, 1918–1921*. Ed. by Michel Janssen et al. Princeton: Princeton University Press.

Eucken, Arnold, ed. (1914). *Die Theorie der Strahlung und der Quanten. Verhandlungen auf einer von E. Solvay einberufenen Zusammenkunft (30. Oktober bis 3. November 1911)*. Abhandlungen der Deutschen Bunsen-Gesellschaft für angewandte physikalische Chemie, 7. Halle an der Saale: Knapp.

Goenner, Hubert, Jürgen Renn, Jim Ritter and Tilman Sauer, eds. (1999). *The Expanding Worlds of General Relativity*. Boston: Birkhäuser.

Kaiser-Wilhelm-Gesellschaft (1922). *Kaiser-Wilhelm-Gesellschaft zur Förderung der Wissenschaften. Jahresbericht April 1921–Oktober 1922*. Burg/Magdeburg: Hopfer.

Kox, Anne J. and Daniel Siegel, eds. (1995). *No Truth Except in the Details*. Dordrecht: Kluwer Academic.

Mayer-Kuckuk, Theo, ed. (1995). *150 Jahre Deutsche Physikalische Gesellschaft. Weinheim: VCH Verlagsgesellschaft*. Special issue of *Physikalische Blätter*, 51 (1995), F 1–F 240.

Renn, Jürgen, ed. (2007). *The Genesis of General Relativity*, 4 vols. Dordrecht: Springer.

Schwabe, Klaus, ed. (1988). *Deutsche Hochschullehrer als Elite 1815–1945*. Boppard am Rhein: Boldt.

Warburg, Emil, Max von Laue, et al. (1918). *Zu Max Plancks sechzigstem Geburtstag*. Karlsruhe: Müllersche Hofbuchhandlung.

© The Author(s), under exclusive license to Springer Nature Switzerland AG 2020
H. Goenner and G. Castagnetti, *Establishing Quantum Physics in Berlin*,
SpringerBriefs in History of Science and Technology,
https://doi.org/10.1007/978-3-030-63122-2

Printed in the United States
by Baker & Taylor

Printed in the United States
By Bookmasters